Transport Issues and Problems in Southeastern Europe

Edited by
CARALAMPO FOCAS

Routledge
Taylor & Francis Group

LONDON AND NEW YORK

First published 2004 by Ashgate Publishing

Reissued 2018 by Routledge
2 Park Square, Milton Park, Abingdon, Oxon OX14 4RN
605 Third Avenue, New York, NY 10017

First issued in paperback 2021

Routledge is an imprint of the Taylor & Francis Group, an informa business

ISBN 13: 978-0-815-39858-5 (hbk)
ISBN 13: 978-1-351-14416-2 (ebk)
ISBN 13: 978-1-138-35779-2 (pbk)

DOI: 10.4324/9781351144162

Contents

List of Figures

List of Tables

List of Contributors

Georgia Aifadopoulou, TransEuropean Consulting Unit of Thessaloniki

Johannes Baur, Commission of the EU, Brussels

Alexandros Deloukas, Attiko Metro. Athens

Günter Emberger, University of Technology, Vienna

Simeon Evtimov, Project Implementation and Management Unit of New Bulgaria-Romania Danube Bridge Project at Vidin-Calafat

Caralampo Focas

Chris Germanacos, Louis Berger Group, Bucharest

George Giannopoulos, Aristotle University of Thessaloniki

Boyan Kavalov, University of the National and World Economy, Sofia

Hermann Knoflacher, University of Technology, Vienna

Markus Mailer, University of Technology, Vienna

Marco Mazzarino, University of Trieste

Vasilis Mizaras, Transport Research Unit of Thessaloniki

Antonín Peltrám, ex-Minister of Transport, Prague

Paul Pfaffenbichler, University of Technology, Vienna

Christos Pyrgidis, Aristotle University of Thessaloniki

Evangelos Sambracos, University of Piraeus

Gi Seog Kong, Technical University of Budapest

Katalin Tánczos, Technical University of Budapest

Péter Rónai, Technical University of Budapest

Zoltán Bokor, Technical University of Budapest

Mateu Turró, European Investment Bank

Foreword

Transport, both as an industry and as a policy area, is an unquestionably important factor in European economic development and has played a particular role in the post-war European integrative process. European goods and people are more mobile than ever before and transport policy has emerged as one of the few dynamic policy areas, whose origins date back to the Treaty of Rome itself. For many years, the primary role of the Common Transport Policy has been to remove the technical and institutional barriers that exist between the member states and this Policy continues to be a major aspect for Europe, and one of the most important agenda points in the discussions with the acceding countries for the adoption of the acquis communautaires.

The Trans-European Network (TEN) has gradually arisen as one of the driving forces for the achievement of growth, competitiveness and employment. In preparation for the enlargement of the European Union to the east, the extension of the existing TEN in the candidate countries for accession comprises an effort to upgrade existing infrastructures or build new ones in order to create a coherent network out of the current patchwork of transport links and achieve safe and speedy connections between countries that are necessary to increase the efficiency of the Single Market and maximise the potential of European trade.

Another part of the European wide transport infrastructure concept, is the Corridor concept, which has developed over the past eight years along with the Pan-European Transport Conferences. During the Conferences in Crete in 1994 and in Helsinki in 1997, the UN-ECE, the ECMT, the European Commission and all European states supported the development of ten pan-European Transport Corridors and four pan-European Transport Areas (PETrAs). These ten multimodal transport Corridors and Areas, provided an important focus for investment by the international financial institutions, and significant progress has been achieved in their development.

Greece's significance for the region of Central and South Eastern Europe is unquestionable. It has already been the focus of many new advances and developments having a special mission for the peace and the prosperity of the area. Greece has excellent relations with all its neighbouring states. It is now feasible for Greek companies, which are already active in several business sectors, to move into the transportation industry in the greater area of Central and Eastern Europe.

In this respect, Greece strongly supports the development of the pan-European Transport Corridors, in fact the Greek Ministry of Transport and Communications has undertaken the Presidency of Corridor X and acts in close cooperation with the other Corridors and particularly Corridors IV and VII.

Apart from the above mentioned initiatives, Greece is currently undertaking multimodal projects all over its territory with the scope of improving substantially the transport infrastructure and quality of the network of the country, thus improving both the transport cohesion with the rest of the Balkans and the EU. The objective shared by the countries of the region is the complete reconstruction, rehabilitation and upgrading of the whole transport infrastructure in Southeastern Europe. And after having taken into account the particularities of Greece, the government has opted for a progressive removal of restrictions in transport operations. It has been judged that the immediate and unconditional liberalisation of the transport sector, as it has been practised in other countries, would create market disruptions with disastrous effects in the case of Greece and the whole region.

All those who are involved in policy making should profit from the opportunity to take all necessary political decisions towards the rapid development of the transport infrastructure and quality of operations in Southeastern Europe, which will be of significant benefit to the entire region.

Concluding, it has to be mentioned that initiatives which contribute to improving the quality of transport operations in the region, such the Southeast Transport Research Forum (SETREF) and this book in particular, are most welcome since they reflect the importance of the development of the region and provide the necessary background for profitable cooperation of the countries in the field of Transport.

Christos Verelis

Chapter 1

What is Missing in Southeastern Europe?

Caralampo Focas

Transport in Southeastern Europe is undergoing a period of transition which in the last 15 years has seen a fall in demand (especially for rail and inland shipping), an increasingly ageing infrastructure and the quality of service of transport provision lagging far behind that of Western Europe. This could not but be so, since there have been recent great political upheavals and wars in the Balkans and major changes of economic structures in the previous command economy states.

This book is based on papers given and discussion that followed in the first international conference 'Cost Effective Infrastructure and Systems to Improve Cargo and Passenger Transport in Southeastern Europe' held in Budapest in October 2001 by the Southeastern Transport Research Forum (SETREF).[1]

The purpose of this introductory chapter is to provide an overview of recent trends, current salient issues and the major problems in transportation in this part of the world. It aims to do this through a critical interpretation of the papers and discussions that took place at the Budapest conference. Finally, this paper will also present the viewpoint of the author on what is missing in the transport field in Southeastern Europe.

An Overview of Transportation in Southeastern Europe

Transportation in Southeastern Europe is characterised by an ageing and often inadequate transport infrastructure and antiquated and out-of-date management systems of transport operations. Furthermore, transportation activities take place in an environment of political and economic instability.

From the 1980s we have seen a dramatic fall in economic production in most of the countries in Southeastern Europe. The exceptions to this have been Greece and Turkey, the latter having followed a quite different, but nonetheless bumpy, path. This great economic plunge has in most cases been related to the collapse of the socialist command economies of the Eastern European countries. In the countries of the former Yugoslavia, Albania and

Moldavia the economic collapse has been compounded by ethnic wars and civil strife. For instance, while in 1989 the gross national product (GNP) per capita in Yugoslavia was over US$3,000, by 1993 it had fallen to under US$1,500. Since then it has hovered around that figure and has not grown again (Depolo, 2001). In many countries of the region there is still a lack of security to guarantee smooth and safe transport movements.

The railways have been the mode of transport most badly affected by these events. In the late 1980s and 1990s there has been a savage decline in rail traffic in most of Southeastern Europe.[2] For instance, in Hungary the share of freight transport by railways has fallen from 70 per cent to 30 per cent between 1989 and 2000.[3] In countries of former Yugoslavia, rail transport now carries a near insignificant amount of traffic. In the Czech Republic the railways carried 25 billion tonne-kilometres of freight in 1993. By 1999, this figure had fallen to 17 billion tonne-kilometres.[4]

Yet car ownership and road traffic in the countries of this region have produced a very different picture with significant growth rates. For instance, in the Czech Republic in 1993 there were 148,000 lorries registered in the republic and this more than doubled by 1999 to over 323,000. Car ownership also grew over this period by over 30 per cent.

The wars in the former Yugoslavia have had the effect of damaging the transport infrastructure, greatly depressing transport demand due to economic retrenchment, and a massive decline in through traffic due to inadequate security en route (Vukanovic et al., 2001). Paradoxically, the collapse of through traffic yet has had the effect of increasing traffic in some neighbouring countries.[5] For instance, although the Bulgarian economy has suffered greatly in the past decade, the country has seen increasing through traffic, from Greece and Turkey to Northern and Western Europe (Boyan, 2001). For the same reason of bypassing Yugoslavia, there has also been an increasing amount of short-sea-shipping, especially in ro-ro (roll on, roll off) traffic on the Ionian and Adriatic seas between Turkey and Greece to Italy and Croatia.

Transport infrastructure in Southeastern Europe is often of poor quality, ill maintained and differs in specification from country to country. The level of service on Southeastern Europe's road and rail systems is variable but most often poor.

Although on paper there are the grandiose pan-European Transport corridors, in practice in Southeastern Europe, at present these often do not mean much more than lines on a map. On the roads, there are numerous problems in terms of bottlenecks, such as small bridges (or nonexistent bridges[6]), poor maintenance and slow vehicles, such as horse drawn carts.[7]

There very few common standards on infrastructure, roads and most importantly, rail operations throughout Southeastern Europe. On the rail side, the problem of differing systems has the effect of making interstate travel problematic, cumbersome and prone to great delay at border crossings. For instance, on trans-European rail corridor IX there are differing gauges, electrification systems, permitted weights and signalling systems, not to mention different standards of operations[8] and therefore no through services.

Transport Infrastructure in Southeastern Europe and the Perceived 'Missing Transport Links'

There is a widespread feeling in the countries of Southeastern European, and a majority of the participants at the SETREF conference, that they are lacking in infrastructure, especially roads and motorways. These are usually labelled as the 'missing transport links' or 'transport gaps'.

A graphic example of this was the announcement that it took 27 hours for some conference attendees to come from Sofia in Bulgaria to Budapest by rail. This was stressed to demonstrate that there is indeed a shortcoming in the transport infrastructure of the Balkans. At the conference, nearly all demands for funding (from the European Union) of transport schemes affected primarily the construction of new roads and, to a smaller extent, the funding of the modernisation of the railways and ports. There were very few demands for schemes to improve the quality of transport, such as traffic management, with few exceptions such as the demand for the establishment of cross-border transportation centres (Kokoritsos, 2001).

The widely used definition of the 'missing transport links' is problematic. There seem to be three different methods of defining the 'missing transport links' by those who use this term in Southeastern Europe.

- The first method is based on subtracting the kilometres of roads, motorways, railways, etc. per capita in the countries of Southeastern Europe from the average of the same measures for Western countries. The 'missing transport links' are not just confined to roads and rail but can be measured in terms of ports, bridges, or tunnels.
- The second method is based on identifying on a map the sections of the pan-European corridors that do not have the infrastructure that is proposed. Often it includes 'bottlenecks' such as missing bridges, but

can also include large sections, which for instance are roads not of closed motorway standard.

- The third method of defining the 'missing transport links' is not based on current traffic demand or even on short-term future demand, but rather on what are termed 'geostrategic' considerations. However there is no defined methodology for:

  - identification of these 'geostrategic' infrastructure projects;
  - prioritisation of these schemes; and
  - assessment of such schemes.

Many transport professionals in Southeastern Europe argue that some 'missing transport links' or 'bottlenecks', such as bridges between Bulgaria and Romania, should not be assessed purely in terms of forecasted demand, but also to maintain a 'geostrategic balance' in the Balkans and 'to serve as a stabilising factor in the region' (Boyan, 2001). It was argued that one should see the historic importance of trade routes in defining 'geostrategic' importance. For instance, it was claimed that pan-European corridor IV is vital to linking Greece with the rest of Europe, but there are still no bridges across the Danube, between Bulgaria and Romania. If the bridges between Bulgaria and Romania are not erected, it is argued that this could prove to be a destabilising element in the region (ibid.).

Another example of 'geostrategic' considerations in establishing a 'missing transport link', mentioned at the Budapest conference, is the railway link on pan-European corridor V between Slovenia and Hungary. It was claimed that it 'serves a strategic role' for the North Atlantic Treaty Organisation (NATO) and this has been an important reason for its completion. A purported justification of this claim has been that over 80 per cent of its funding came from Germany (DM120 million).[9]

As these examples show, the definitions of 'missing transport links' are not exclusive and can be a combination of the above.

The great enthusiasm in political and construction circles for the adoption of these corridors and the building of motorways to meet these so called gaps, is often leading to biased and unrealistic forecasts. There are many examples of built infrastructures where the forecasted demand has not materialised. A typical example of this is motorway M1 in Hungary. After the M1 was opened in January 1996, traffic and revenues were below projection, because most Hungarian motorists elected to take slower, parallel non-tolled routes. The failure of the revenue and traffic to materialise led the concession company to be taken over by the Hungarian government in June 1999.[10]

Another sorry example is the Yugoslavian motorways. The level of internal demand is not enough to repay the loans it cost to build these motorways, so according to Vukanovic, one of the speakers at the Budapest conference, it is important for Yugoslavia to attract through traffic, solely for the purpose of creating revenues to repay the loans for this infrastructure. Ten years ago Yugoslavia had 10,000 vehicles a month in through traffic. This has fallen to about 1,000 vehicles per month (Vukanovic, 2001).

The Preoccupation with Trans-European Corridors in Southeastern Europe

There has been an over-concentration on and over-preoccupation with infrastructure networks in Southeastern Europe. Much less, if any, attention is paid to efficient operations and product quality.

The over-concentration on infrastructure has been exacerbated by a series of very influential studies and policies concentrating on defining a road and rail network for the whole of Southeastern Europe such as the pan-European corridors, the Transport Infrastructure Needs Assessment (TINA), and more recently the Transport Infrastructure Regional Study (TIRS). This type of identification of vital 'corridors' and 'networks' as lines on a map, aids the notions of 'missing transport links'.

The initial plan for a multimodal European transport network was, the Trans-European Network for Transport (TEN-T).[11] This was the network of key transport infrastructure identified across the 15 European Union states in 1996, which was amended in 2001.[12] The TEN-T guidelines encompassed a list of 14 priority projects endorsed by the European Council in 1994. The only priority project that was to be undertaken in Southeastern Europe was the completion, upgrading and construction of new motorways in Greece. For the TEN-T, the European Union has earmarked about €4 billion, which excludes other funding obtainable through the cohesion funds and special bank loans.

The setting of the trans-European network was followed by the Transport Infrastructure Needs Assessment (TINA). This was an initiative set up in 1996 by the European Commission to identify the scale of the problem for creating an effective European transport network across the majority of the Central and Eastern European Countries of Europe (CEEC). It was composed of the candidate countries for accession to the European Union, with the exception of Turkey and Malta.

A European-wide transport ministers conference held in Crete in 1994 identified nine long-distance transport corridors for development outside the European Union. A later conference at Helsinki established a tenth corridor and also proposed four, somewhat nebulous, pan-European Transport Areas. These trans-European network corridors, now known as either the Crete or the Helsinki corridors, have created a transport network that stretches from the Baltic to the Adriatic and Black seas across Central and Eastern Europe and into the adjoining countries of the Commonwealth of Independent States (former Union of Socialist Soviet Republics). They comprise road, rail and some inland waterways. One corridor, corridor VII, is different in that it is the River Danube, running from the Black Sea along the borders of the Ukraine, Romania, Bulgaria, across Serbia and Hungary into Slovakia.

Investment in 'corridor' infrastructure may not accord with the priorities for infrastructure investment identified by national governments elsewhere on their respective transport networks. The need for these new roads and railways is certainly not proven. They may not accord with national priorities and indeed may cater more for through traffic. Furthermore, most often there are no state funds to build major new transport infrastructure and private sector involvement in building infrastructure is problematic, because of political uncertainty, low traffic projections and inadequate government guarantees.[13]

There is no legal basis for the pan-European corridors (not all Memoranda defining the actual routing of corridors are formally established); even less so for the TINA networks. Yet they have established some sort of mythical status in Southeastern Europe. An often-heard pronouncement for new road construction in these countries, is to 'complete the corridor' or 'it's part of the TINA network'.

The latest developments in the definition of a comprehensive transport network in Europe have been a number of studies, including the Transport Infrastructure Regional Study (TIRS).[14] This has been an initiative to complement the work of TINA but specifically aimed at the Balkan countries of Albania, Bosnia Herzegovina, Bulgaria, Croatia, FYR of Macedonia, Romania and Yugoslavia. The majority of these countries had been omitted from the detailed studies relating to TINA.

Although European Union transport policy favours rail and traffic management, this is often overlooked in Southeastern Europe where there is a strong desire to build motorways.

Short Sea Shipping in Southeastern Europe

The preoccupation with the pan-European Network corridors ignores a major form of transport in Southeastern Europe, short sea shipping. About 70 per cent or more of tonne-kilometres of freight traffic in intra-European transport takes place by short sea shipping.[15] Recently more attention has been paid to short sea shipping by the European Commission with the adoption of the concept of the 'Motorways of the Sea', an extension of the TEN in the sea. Four such Motorways of the Sea have been proposed and by 2004 they were at the phase of being geographically defined. Southeastern Europe has one such sea corridor, commencing from Slovenia, down the Adriatic Sea into the Ionian Sea and ending in Cyprus. However, the concept of the trans-European corridors still ill suits the geography of parts of Southeastern Europe which is comprised of numerous small scattered islands, principally in the eastern Adriatic Sea, the Agean Sea and eastern Ionian Sea. These islands cannot be linked up by a single corridor or a Motorway of the Sea.

Shipping is important in Southeastern Europe both in terms of short sea shipping, since it has a long coastline with numerous port cities, and inland shipping with Europe's longest navigable river, the Danube, which crosses or borders ten countries.

Overall we have seen a fall in traffic and goods volumes from the late 1980s both for short sea shipping as well as for inland shipping, especially in the Black Sea (Breshkov, 2001). The level of traffic on the Danube before the most recent war in Yugoslavia (1998) had already fallen to 43 per cent of the level it was in 1989 (Vukanovic, 2001). Following the NATO bombardments of Yugoslavia, freight in the Danube has become very limited; for example all the bridges at Novi Sad were destroyed. In contrast, the overall level of short sea shipping between Greece and Italy has grown, largely due to the decline of road-based transport through the former Yugoslavia due to the recent wars and ongoing political instability and lack of security.

Quality (Lack of It)

The transport policies of the European Union are pressing both for an increase in quality, especially the reduction of emissions, through the White Paper on European Transport Policy for 2010 and at the same time also call for the completion of the trans-European transport network.[16]

The issue of a lack of quality in transport provision in Southeastern Europe is not a major concern amongst most local transport practitioners and politicians. Yet occasionally in the literature and at conferences one reads and hears about examples which mention where it is lacking and measures which need to be taken to improve it.

There is a widespread view in the Southern and Eastern European countries that there is a lack of basic transport infrastructure, especially roads, and that it is imperative, as mentioned earlier, that the perceived gap between Southern and Eastern European countries and the countries of Northern Europe be bridged. The quality of the infrastructure is often not considered an important issue, nor the upgrading of the quality of existing transport modes. European Union funds are sought to build new large projects, mostly, but not exclusively, roads.

These views miss out completely the disadvantages that are caused by the poor quality of existing infrastructure, its inadequate maintenance and the low level of transport and traffic management and inadequate information systems. This low level of service not only has a negative impact on the quality of life of the residents of these regions but also a high economic cost, created by low productivity and unreliability of the transport system.

There are numerous examples of poor quality of infrastructure that could be analysed; however, a very particular example chosen for this book, that was presented at the Budapest conference, is that of the Corinth Canal,[17] a major transport route that saves many hours for ships travelling between Italian ports and those of the Levant. The canal currently has no facilities such as shops, banks or any information services. It possesses only an administration office to process the ships that want to cross. The lack of service quality can been seen throughout Southeastern Europe on all modes of transport, from run-down rail services to badly constructed and maintained highways.

Yet there are some signs that times are changing. There are some few projects to improve quality in transport in the countries of Southeastern Europe, as mentioned at the SETREF conference. One of these is the development of a decision support system for internal shipping in Greece (Lekakou and Stergiopoulos, 2001). The reason for developing such a tool was the recognition of the poor regulatory system of the Greek Ministry of Merchant Marine both in terms of determining schedules, fares structures, but also that of safety. The tragedy of the sinking of the Samina ferry boat in 2000, with great loss of life, brought to the fore the neglect of proper regulation, adequate quality and adherence to safety procedures.

However, the preoccupation with improving service quality is not widely diffused in the Balkans. The examples of projects associated with improving transport quality reported at the conference in Budapest centred entirely on Greece, a member of the European Union.[18]

There are also minority voices that criticise the entire process or ideology that is leading for a push in creating extra transport infrastructure instead of concentrating on improving the quality of the habitat in these regions. Standard transport planning methodologies that rely on traffic forecasts, improved road speeds and concomitant cost-benefit analyses that lead to more and bigger roads are being severely questioned.[19] It has been argued that funders of transport schemes should be funding particular outcomes (goals) rather than projects. This could lead funding liveable cities rather than new roads; and furthermore, Southeastern European cities now have the chance of creating a better quality of life than the cities of Western Europe which have copied the car-dominated North American paradigm.[20]

Conclusions

This book provides a snapshot of the preoccupations and debates regarding transportation in Southeastern Europe at the beginning of the third millennium. This region has seen a great decline in rail and inland shipping due to the collapse of the economies of the Eastern Europe following the collapse of the Soviet Union and the ethnic wars that have ravaged many Balkan countries.

There is a great preoccupation in Southeastern Europe with transport infrastructure networks. This has taken form in particular in terms of the pan-European corridors. It is perceived that these corridors must be completed and concomitantly the transport 'missing links' be filled in.

Much less, if any, attention is paid to efficient operations and product quality of the transportation system. So although the transport infrastructure is increasingly ageing, the quality of service of transport provision is even further lacking. There are very few projects to increase the quality of passenger transport compared to projects to create new infrastructure, primarily to build new roads.

This book presents papers that make the case for new infrastructure, as well as papers that question the notion that new infrastructure is what the region needs. These papers, whether they advocate more quantity or quality are embedded in the local realities of the countries in region.

Notes

1 This paper has also been informed by the debates that took place at the 15th International Symposium on Theory and Practice in Transport Economics by the European Conference of Ministers of Transport which was held in Thessaloniki in June 2000. A full account of the debate of this conference can be found in Focas, 2002.

Furthermore, this book includes a specially commissioned chapter, 'Organisation, Monitoring, and Efficiency of the Athens Urban Public Transport', from Alexandros Deloukas (Chapter 10) to enhance the debate on the 'quality in transport'.

2 See Chapter 3, 'Transportation Corridors for Europe and their Development', by Chris Germanacos.

3 See Chapter 7, 'Major Improvements and Challenges in Transport Logistics along Corridor V', by Péter Rónai and Katalin Tánczos.

4 See Chapter 5, 'The Czech Transport Infrastructure and Pan-European Corridors', by Antonín Peltrám.

5 See Chapter 5.

6 See Chapter 8, 'Construction of a New Combined Road-Rail Bridge over the Danube at Vidin-Calafat and its European Integration Potential', by Simeon Evtimov.

7 See Chapter 3.

8 See Chapter 6, 'The Prospects for the Development of the Southeast Section of Rail Corridor IX', by Christos Pyrgidis and George Giannopoulos.

9 In Chapter 7, Rónai and Tánczos make the point that in 1980 the railway tracks between Slovenia and Hungary were ripped up. Now the link has been re-established.

10 See Chapter 15, 'Government Policy to Facilitate Private Participation in Toll Road Infrastructure through Investment Mitigation Measures Focused on Eastern Europe and some Asian Countries', by Gi Seog Kong and Katalin Tánczos.

11 Decision No. 1692/96/EC.

12 By early 2004 the TEN was under revision.

13 See Chapter 3.

14 See Chapter 3.

15 See Chapter 9, 'The Role of the Corinth Canal in the Development of the Southeastern European Short Sea Shipping', by Evanghelos Sambrakos.

16 See Chapter 2, 'European Union Transport Policy and Enlargement', by Johannes Baur.

17 See Chapter 9.

18 These were: i) the construction of cross-border transport centre for freight distribution at Promachonas on the Greek-Bulgarian border (Kokoritsos, 2001); ii) smartcard development for paying public transport and road tolls in Thessaloniki (as can be seen in Chapter 11); and iii) the creation of computerised incident detection system of the ring-road of the same city (Nathanail, 2001).

19 See Chapter 16, 'Considering Multimodal Capacity in the Assessment of Road Design', by Markus Mailer. One of the main criticisms made is that the 'benefits' are composed of nonexistent 'time-savings'. See also Chapter 17, 'Sustainable Cities for Countries in Transition – Learning from Mistakes in Countries with High Motorisation' by Hermann Knoflacher.

20 See Chapter 17.

References

Boyan, K. (2001), 'Main Problems of Dry Cargoes Handling in Major Bulgarian Ports', paper presented to the conference 'Cost Effective Infrastructure and Systems to Improve Cargo and Passenger Transport in Southeastern Europe', Southeastern Transport Research Forum, Thessaloniki.

Breshkov, I. (2001), 'Forecasting the Transport Flows in Black Sea Region', paper presented to the conference 'Cost Effective Infrastructure and Systems to Improve Cargo and Passenger Transport in Southeastern Europe', Southeastern Transport Research Forum, Thessaloniki.

Depolo, V. (2001), 'Belgrade's Public Transport System on the Way towards Economic and Institutional Changes – Is a Spontaneous Transition Possible?', paper presented to the conference 'Cost Effective Infrastructure and Systems to Improve Cargo and Passenger Transport in Southeastern Europe', Southeastern Transport Research Forum, Thessaloniki.

Focas, C. (2002), 'Report of the 15th International Symposium on Theory and Practice in Transport Economics', in 'Key Issues for Transport Beyond 2000', European Conference of Ministers of Transport, Paris.

Kokoritsos, P. (2001), 'Private Sector Involvement in the Transport Sector. The Case of Cross-border Transportation Centres', paper presented to the conference 'Cost Effective Infrastructure and Systems to Improve Cargo and Passenger Transport in Southeastern Europe', Southeastern Transport Research Forum, Thessaloniki.

Lekakou, M. and Stergiopoulos, G. (2001), 'Maritime Transport Organisation in Greece: A Decision Support Tool', paper presented to the conference 'Cost Effective Infrastructure and Systems to Improve Cargo and Passenger Transport in Southeastern Europe', Southeastern Transport Research Forum, Thessaloniki.

Nathanail, T. (2001), 'A Highway Incident Management and Traffic Control System based on Video Image Processing', paper presented to the conference 'Cost Effective Infrastructure and Systems to Improve Cargo and Passenger Transport in Southeastern Europe', Southeastern Transport Research Forum, Thessaloniki.

Vukanovic, Sm., Jovic, J. and Tubic, V.I. (2001), 'Traffic Flow Characteristics in South-Eastern Europe – Present state and Perspectives', paper presented to the conference 'Cost Effective Infrastructure and Systems to Improve Cargo and Passenger Transport in Southeastern Europe', Southeastern Transport Research Forum, Thessaloniki.

PART I
EUROPEAN POLICY AND
TRANSPORT CORRIDORS

Chapter 2

European Union Transport Policy and Enlargement[1]

Johannes Baur

Enlargement and the European Union Transport Policy

The enlargement of the European Union is a great challenge for European Union transport policy. It will lead to an extension of the transport policies, including the Trans-European Network for Transport (TEN-T), to the Central and Eastern European countries (CEEC) through the alignment with European Union legislation in the Accession Countries.

Inside the European Union distances between East and West will become greater, and a country like Ukraine will become an important neighbour and trade partner with 50 million inhabitants. Enlargement is not in the distant future, but will happen soon. The European Union has demonstrated its clear commitment to moving the accession process forward and to concluding the negotiations with the most advanced countries in 2002, if the preparations and the negotiations continue at the same pace as in 2001. The first countries may thus join the European Union before the elections to the European Parliament in 2004.

Transport is a key element for the functioning of our society and our economy. Transport not only contributes to the exchange of goods, but is also an essential prerequisite to realise the project of an integrated Europe.

Inside the European Union, we can state a contradiction between the growing need for individual mobility and a public opinion which becomes more and more intolerant of chronic delays, bad service and environmental pollution.

Despite the successful development of the transport market in the European Union, there are a series of problems to which we have to offer a political reaction:

1   the unequal growth in different modes of transport;
2   congestion on the main road and rail routes, in towns and at airports;

3   harmful effects on environment and public health.

I would like to present the main guidelines of the White Paper and then analyse the consequences of its basic requirements for Southeastern and Eastern Europe. I would emphasise in particular the following challenges for Transport Policy in an enlarged Europe until 2010:

- an overall task is to include transport policy in the European Union Strategy for sustainable development. This means decoupling transport growth and GDP growth by the shift from road to rail, water and public passenger transport. Today, 79 per cent of passenger transport and 44 per cent of goods transport is carried by the road. If nothing is done, CO_2 emissions from all transport should increase by 40 per cent by 2010. Transport will have to contribute to reduce emissions in order to reach the targets of the Kyoto Protocol. Figures show that CO_2 emissions due to road transport accounts alone for over 80 per cent of all transport emissions;
- enlargement will contribute to the growth of transport needs in the next decades. The growing need for individual mobility and the transition to an global economy will lead to 38 per cent more goods transport and 24 per cent more passenger transport in Europe. If the current development continues, transport of heavy vehicles will increase by 50 per cent in 2010. In view of the already existing problems of congestion in Europe, this would hardly be tolerable. Already 10 per cent of the European road network suffers from chronic congestion daily, and 20 per cent of the rail network can be considered as problematic.

Together with enlargement, the obvious need for a policy promoting sustainable development offers an opportunity to adapt the European Transport Policy. The European Commission has reacted by publishing a new White Paper on European Transport Policy for 2010. The Commission emphasises the need to make other modes of transport more attractive and proposes to introduce a tariffication which accurately reflects the internal and external costs of every mode of transport. The European Council of.Göteborg has called for a shift of balance between transport modes, basically from road to rail, inland waterways and short sea shipping. The objective of the Commission is to stabilise in 2010 the shares of transport modes at the level of 1998. This implies reducing the growth of road transport, whereas rail and inland waterways should triple their growth figures. In our view, this is both a realistic and an ambitious objective. The example of Japan (passenger) and USA (freight transport) shows that railways can operate successfully in developed societies.

How can we reach these objectives?

- Enlargement is an opportunity for railways: the importance of railways in Eastern Europe is higher than in Western Europe (40 per cent as against 8 per cent). The Commission's aim is to revitalise railways and to create a new balance of the different modes of transport. Railways could play an essential role in solving transport problems: less pollution, less congestion and fewer accidents. Thus, the revival of railways is one of the priorities of the European Union transport policy. The Commission is pushing the liberalisation of the railways market in the Union, which is a necessary step towards making international freight transport more competitive. Moreover, the technical harmonisation of the different national railways in Europe is an important task. To implement these goals, the European Union has adopted in early 2001 an important railway package. What is the key content of this new railway legislation?
  - A first objective is the opening of the European railway market, which is still split into different national sectors. From 2003, the trans-European rail freight network, with about 50,000 km, will be opened. This network will also be extended to the new members of the European Union.
  - To ensure the access to the network, the European Union has defined clear rules for the allocation of capacity including the separation of an infrastructure manager, responsible for the network, and railway undertakings.

Other objectives are:
  - the promotion of combined transport is an important task for the European Commission to reduce CO_2 emissions in the transport sector;
  - to improve road transport conditions in Europe by making the roads safer and reducing greenhouse gas emissions. Road transport has a high number of deaths and fatalities (about 41,000 in the European Union) in comparison to other transport modes;
  - in general, the existing infrastructure has to be improved and bottlenecks have to be eliminated. Insufficient networks in the candidate countries and congestion all over the European Union may seriously affect our economy. If nothing is done, the costs of congestion will grow to 1 per cent of the European Union GDP. The trans-European network will be

adapted to the needs of the new member states and to the increasing transport needs from East to West and from Southeastern to Central Europe.

- On maritime transport, the situation will also change considerably. With the accession of Malta and Cyprus the European Union will nearly double its merchant fleet. Romania and Turkey have also important fleets. In the accession negotiations, the European Union is emphasising the need to enhance safety in maritime transport in Europe and to promote short-sea shipping. The European Union has adopted an important package of new legislation on maritime safety – the so-called 'Erika packages'. Since European Union member states and candidate countries like Romania and Bulgaria will have to implement these measures, maritime safety in the Black Sea and in the Mediterranean will considerably improve.

In all fields of transport, candidate countries have made good progress in adopting the necessary legislation. However, there is still much to be done and particular efforts have to be made to implement and enforce the European Union standards.

Thus, the main challenges for transport policies of the European Union in the context of the enlargement will be:

- to strengthen the still strong position of railways in the freight market of candidate countries by restructuring and modernising railway companies and opening the market. It is the goal of the European Union to maintain the modal share of rail freight transport in the candidate countries in 2010 at 35 per cent;
- to reduce infrastructure bottlenecks and to include the candidate countries' international network into the trans-European network. Private capital will have to play an important role there;
- the implementation and enforcement of European Union standards in the accession countries. This is especially important with regard to the forthcoming *acquis* in maritime transport (Erika packages). To cope with this task, the European Union will continue to assist the future member states to build up the administrative capacity in the transport sector notably by training inspectors and administrative staff responsible for enforcing transport legislation.

Enlargement and Eastern Europe

Trade and transport in Eastern Europe will profit from European Union enlargement. Economic growth and raising investment in infrastructure (including European Union funds before and after accession) will increase the importance of transport.

The liberalisation of European Union railways may also stimulate changes for the railway policy of other countries. The modernisation of European railways should be a priority not only for the enlarged Union, but also for other countries. The European Union's goal to revive railways by opening the market and strengthening competition could be an example for other countries. Oversized and inflexible national railway operators will not be able to compete with other modes of transport, notably road transport.

The restructuring of railways in candidate countries and in Eastern and Southeastern Europe is not an easy task. Rolling stock is mostly outdated, and the railway operators do not have the money to invest in infrastructure. As to passenger transport, wagons lack modern equipment and are uncomfortable. Trains have to be updated in order to attract passengers. On rail freight transport, the service will also have to be improved. Even more important is the aspect of time: to be competitive, railway operators will have to offer high-speed connections between the main cities of a country. The modernisation of management structures is another difficult task, since they have to take place in companies with a strong link to the state and with a high number of employees.

Nevertheless, things are moving.

However, railways in Central and Eastern Europe have great potential. Their position in the competition with road transport is still much better than in the existing European Union. The 40 per cent market share will be a good starting position. Many Central and East European railways have already started modernisation and restructuring.

For international transport operators from Eastern Europe to the European Union, there will not be many changes. Most technical and social requirements in international road transport between the European Union and third countries are covered by international agreements. All hauliers transporting goods to the European Union should already conform to these provisions. The implementation of the European Union requirements in the accession countries will have consequences for intra-European Union transport rather than international transport. For road transport, this embraces the fitting of tachographs and speed limitation devices, weights and dimensions, fiscal harmonisation, roadworthiness tests, access to the profession, access to the market for operators, etc.

Nevertheless, the Commission attaches great importance to the harmonisation of European transport legislation with the aim to enhance international transport operations and the development of a sustainable transport market. Eastern European countries will benefit from introducing European Union safety rules for road safety and the transport of dangerous goods or for weight and dimensions.

Eastern European countries are already participating in the European Union's 5th Framework Programme for Research and Development. Sustainable mobility and land transport and marine technologies with a budget of about €690 million are important priorities in the field of transport. At the time of writing, the European Union is calling for new proposals with a view to improving the integration of candidate countries. As to the 6th Framework Programme, the Commission has made the proposal to create a European Research Area. Notwithstanding the final decisions in 2002, transport research activities will be included in several priorities of the Programme: information society and technologies with the aim to support vehicle infrastructure and portable systems for integrated safety; aeronautics and space; and in the context of sustainable transport and clean use of energy with the aim to promote clean technologies for transport as well as interoperability and intermodality.

The European Union does not intend to create new barriers between the enlarged European Union and Eastern Europe. Enlargement will be an opportunity to intensify and to validate the close relationships, for example, between Poland and the Ukraine, including the border regions.

The example of the Finnish-Russian border has shown that the introduction of European rules at this border, including the introduction of the Schengen Visa, does not necessarily hamper transport – on the contrary. Today, the borders between accession countries and other Eastern European countries are difficult to cross: regular queues for local and international traffic and long waiting times are often caused by bureaucratic complication and the lack of modern equipment – for passenger transport as well as for goods transport. Having in mind the current problems I think that some fears of negative consequences of enlargement are one-sided and overly pessimistic.

The European Union's PHARE (Poland and Hungary Assistance for the Reconstruction of the Economy) and ISPA (Instrument for Structural Policies Pre-accession) projects in the candidate countries, as well as TACIS (Technical Assistance to the CIS) projects in East European countries, have already contributed to the modernisation of border crossing points. From

1996 to 2000, TACIS has already contributed €58 million to investments on border crossing in Eastern Europe. The European Union has already heavily invested in infrastructure in accession countries. In the framework of the ISPA instrument from 2000 to 2006, around €7 billion was allocated for the 10 acceding countries of Central Europe, dedicated 50 per cent to environmental infrastructure and 50 per cent to transport infrastructure which promotes sustainable mobility, in particular rail and inland waterways.

European Union investments will continue to focus on border crossing and eliminating bottlenecks between accession countries and Eastern Europe: for example, projects at the Polish-Ukrainian and Hungarian-Ukrainian borders. Another goal of these projects is to support the effective operation of border services responsible for border controls. This will include also railway border crossing, where the need for investment to improve the technical difficulties is especially urgent. The most time-consuming procedure for railways at the western borders of Ukraine and Belarus is caused by the different gauge systems, which involves change of bogies for passenger transport and transhipment of goods for freight. Overall, a lot of work to upgrade border crossing points and the hinterland connections to facilitate international transport still needs to be done.

Another important issue will be the further improvement of infrastructure in Eastern Europe. The creation of the pan-European corridors has paved the way for coordinated international actions in this respect. The assessment of the infrastructure needs in this area should be done in a way that the links to the main international network in the candidate countries are clearly established. However, it is in the interests of the Eastern European countries to create the necessary economic and political environment, thus making transport conditions safe and efficient. Investments alone will not suffice. Legislative harmonisation as well as improvement of the administrative capacity and the training of staff should be clearly a priority for all Eastern European countries.

Finally, TRACECA (Transport Corridor Europe Caucasus Asia) has been a key instrument for developing transport infrastructure in the Black Sea region as well as the regulatory framework, training and catalyst investments from international financing institutions (IFIs). It has provided the European Union with a leading role in the region as regards transport policy discussions. Emphasis is now given to the consolidation of the corridor, including the implementation of the Multilateral Agreement on Transport, strengthening the legal framework, improving transport security and consideration of environmental aspects. On the European part of the Black Sea, TRACECA could benefit from an enhanced cooperation with Black Sea pan-European Transport Area and the pan-European Transport corridor IX.

It will need the efforts of all countries in the region and the European Union to cope with these challenges and to make enlargement a success, not only for the European Union and the candidate countries, but also for the whole of Eastern and Southeastern Europe.

Note

1   The views expressed in this paper are personal and do not necessarily reflect those of the European Commission.

Chapter 3

Transportation Corridors for Europe and their Development

Chris Germanacos

Introduction

The purpose of this chapter is to give an insight into the identity and development of transportation corridors in Europe, distinguishing the trans-European network within the 15 member states of the European Union from the 10 pan-European transport corridors ('Helsinki' corridors) which connect with and extend the transport corridors eastwards through the Central Eastern European countries to countries of the Commonwealth of Independent States.

Some of the terminology associated with the corridors and background information is given describing the modal pattern of freight transport throughout Eastern Europe. Maps are introduced to show the pan-European corridors and their relationship to the trans-European network. The extent of the infrastructure associated with these corridors and those of the trans-European network is presented to show comparisons between road, rail and water transport infrastructure.

In examining the task facing Europe for developing these corridors, reference is made to the concepts of improving infrastructure as part of the preparations for increasing membership of the European Union, on the one hand, and how development is being achieved for those not included at present in the accession list, on the other. Objectives and standards for development of infrastructure are identified. Some of the most important initiatives already taken for identifying the scale of the investment required on the pan-European corridors, to meet the current and future needs of transport on the corridors, are described. The potential extent of the pan-European network of road, rail and inland water infrastructure is identified together with estimates of the total cost of investment required for developing the corridors to acceptable levels of service. The financing of these investment costs is discussed.

The concluding remarks focus on some of the difficulties and risks associated with development of the corridors and some recent achievements. Illustrations are taken from Louis Berger's work on a number of the corridors.

Terminology and Background

A number of acronyms have been devised and adopted by European institutions for simplifying the identification of extensive terminology, which is in frequent use today. For ease of reference those acronyms that are used throughout this chapter are summarised below:

CEEC: the Central and Eastern European countries.

CIS: the Commonwealth of Independent States, the countries which lie further east of the CEEC countries.

EU: the European Union of 15 member states.

EC: the European Commission, being the executive charged with the administration of the budget of the European Union and which essentially draws up the ground rules for how Europe functions.

PHARE: originally stood for 'Poland and Hungary Assistance for the Reconstruction of the Economy', Poland and Hungary being the first two countries to emerge from the communist block of Eastern Europe. For more than 10 years PHARE has been the financial and technical support programme for the 13 countries of the CEEC.

TACIS: Technical Assistance to the CIS – a similar programme to PHARE but directed at CIS countries.

TEN: trans-European network. The network of key transport infrastructure identified across the 15 member states. It consists principally of the TEM and TER, which are the motorway and rail routes respectively across the European Union, designated for priority development.

TINA: Transport Infrastructure Needs Assessment. An initiative set up in 1996 by the EC to identify the scale of the problem for creating an effective pan-European transport network across the majority of the CEEC.

TIRS: Transport Infrastructure Regional Study undertaken by Louis Berger Group. An initiative to complement the work of TINA but specifically aimed at the Balkan countries of Albania, Bosnia Herzegovina, Bulgaria, Croatia, Former Yugoslavia Republic of Macedonia (FYROM), Former Republic of Yugoslavia (FRY), and Romania, the majority of whom had been omitted from the detailed studies relating to TINA.

The pattern of freight distribution has been changing progressively in recent years with a significant shift away from rail to road. In a PHARE Multi-country Transport Programme newsletter (European Commission, 2000a), information was given which showed the dramatic effects on freight transport and the respective shares carried by road, rail, pipe lines and inland waterways following the collapse of communism throughout the CEEC from the late 1980s onwards.

It is significant that the share of freight carried by road continues to increase at the expense of declining rail usage. Inland waterways have continued to retain their share. More recent work undertaken by consultants for the TINA initiative (TINA Final Report, 1999) suggests a similar pattern can be expected in the foreseeable future (up to 2015), whatever assumptions are made regarding economic growth and improvements to transport infrastructure. This applies to all sources of freight on the pan-European corridors, whether for the domestic market or that designated for export or import. It should be noted however that the consequences of the bombing of key infrastructure along the Danube during the 1999 war in Kosovo are not reflected in these findings and have certainly led to a reduction in the proportion of freight carried by inland waterways in the region, as evidenced by the current (2001) studies being undertaken for TIRS.

The Transport Corridors of Europe

The pan-European Transport Conference held in Crete in 1994 identified nine long-distance transport corridors for development. A later conference at Helsinki established a tenth corridor. These corridors, now known as the Helsinki corridors, have created a transport network that stretches from the Baltic to the Adriatic and Black seas across Central and Eastern Europe and into the adjoining countries of the CIS. They comprise road, rail and inland waterways; one corridor alone, corridor VII, is the River Danube, running from the Black Sea along the borders of the Ukraine, Romania, Bulgaria, across Serbia and Hungary into Slovakia.

In turn, these corridors link with corresponding nodes of the trans-European network at the borders with the European Union member states. The broad bands of the corridors in reality become specific routes, after discussion with the respective countries along each corridor and subsequent Memoranda of Understanding. The corridors do not necessarily consist of unique routes for road and rail mode but may consist of a backbone route for each mode, supplemented by a number of branches.

Mapping which identifies the pan-European Transport Network and its links to TEN is currently being produced for the EC.[1] This mapping essentially shows the agreed routes which have subsequently been studied in the TINA process and are currently under study by TIRS. Distinctions have been made at this stage in the cartography between those countries, which have European Union pre-accession status and were fully documented through TINA[2] (10 countries of the CEEC) and the remaining countries such as those that have now been included in the TIRS initiative in Southeastern Europe (Albania, Bosnia Herzegovina, Croatia, FRY and FYROM) and those in the CIS (Ukraine, Belarus and Russia).

There is no legal basis for these corridors. This may in part account for some of the difficulties that are being experienced for the comprehensive development of this transport network. Even in mid-2001, not all Memoranda defining the actual routing of corridors are formally established. Furthermore desirable investment in corridor infrastructure to make it a more attractive inter-regional transport facility, may not accord with the priorities for infrastructure investment identified by national governments elsewhere on their respective transport networks.

Table 3.1 below shows the infrastructure lengths of the three main modes of transport, namely road, rail and inland water. Airports and sea ports are also included in the overall Programme for Transport infrastructure development for both TEN and the Helsinki corridor work identified through TINA, but account for a relatively small proportion of the future development budget when compared with these three main modes.

Table 3.1   Extent of infrastructure by mode (km)

Mode	TEN	Pan-European corridors
Road	75,000	13,000
Rail	75,000	14,000
Inland waterways	20,000	2,300

For TEN, the figures in the table above are the extent of the priority network that is attracting private partnership initiatives and national investment supported by structural funds from the European budget.

Examples of development of TEN include the complete redevelopment of the West Coast Main Line railway from London to Glasgow and Edinburgh, currently in the implementation phase. Projects already completed include the Channel Tunnel link between France and the United Kingdom and the associated high speed rail links between Paris, Brussels, Cologne, Frankfurt, Amsterdam and London. More recently the 16 km-long Øresund multimodal fixed link has joined Sweden and Denmark for the first time.

For the pan-European corridors, the corresponding lengths of road, rail and waterways amount to nearly 30,000 km of infrastructure. This is the backbone network of the Helsinki corridors and represents the basic network described by the TINA process. Later it will be shown that this basic network has been expanded by nearly 60 per cent with the addition of further links agreed with the respective countries along the corridors, through the development process and with the inclusion of lengths of corridor which have been omitted from the backbone network to date for political reasons, such as corridor sections in Croatia and FRY.

Whilst the TEN consists of a high speed/high capacity network of transport infrastructure, that for the pan-European corridors is currently very varied. In Slovenia, for example, major investments have been made along parts of the main route of corridor V, constructing tolled motorways equipped with smart card lanes at toll stations for uninterrupted passage. Other motorways on corridor V in Hungary have adopted the Vignette system. However the picture is very different on other sections of the Helsinki corridors where roads may be two-, or at best, multi-lane highways. Railways may be double or single line track, not all with electrification. Adjoining sections of highway on the same corridor may offer significantly different levels of service; tolled motorway facilities may adjoin substandard two-lane carriageway sections winding through tortuous terrain, yet funding is unavailable at present for upgrading the remaining sections to provide a uniform level of service throughout a length of corridor in a particular country. It is little wonder, therefore, that alternative routes are sought along other branches of a corridor in adjoining countries where the highway network may have been more extensively developed.

The Development Challenge

The Task

It is this enormous variety of infrastructure, coupled with the individual perspective and development priorities of each country that has applied to join the European Union, that poses a considerable challenge to Europe in developing the pan-European corridors into an effective transportation network.

Objectives and Standards

In 1997, the then Commissioner for Transport, Neil Kinnock, saw that the development of the Helsinki corridors through a partnership between all the countries of Europe, was a 'very practical way of bringing Europe together again after long years of separation' (European Commission, 1997). Other politically stated objectives include 'rebuilding Europe, united by trade and economic development' (European Commission, 1999a.

Transport-focused objectives, however, are aimed at achieving sustainable mobility, balancing economic efficiency, safety and minimal environmental damage. Particular emphasis has been given to the development of combined transport facilities.

Harmonisation of standards is to be encouraged and indeed is a prerequisite for those countries identified for the European Union enlargement programme. The difficulty with this is that there are no common European Union standards for infrastructure as such. Thus there is often reference by the EC to European Union standards, but in practice, while there is increasing progress towards the development of common product and technical standards, each country tends to have its own references, although similarities may exist from one country to another. The closest Europe comes to harmonisation is through the guidelines published for the TER and TEM and it is generally these that are the reference source for development standards for transport infrastructure, particularly in those pre-accession countries for the enlargement of Europe. These guidelines incorporate geometric design recommendations and the consequences of the various 'Directives' relating to transport and published by the EC, such as maximum vehicle loading and axle weights and environmental emissions.

Network Assessment

The work of TINA in assessing the network has in the author's view made the most significant contribution to the development of the Helsinki corridors to date. It was important for two reasons:

1   it has identified the *global* development requirements of the corridor; and

2   it has identified the potential timeframes for development, given certain
     investment constraints.

For the first time a common approach has been applied to infrastructure
development. In future, investment studies and individual project assessments
must be referenced to much of the work resulting from TINA such as traffic
forecasts, vehicle operating costs and design and construction costs, as well
as the methodology adopted through the TINA process.

The complementary work of TIRS, for the countries that had been earlier
excluded from the detailed work of the TINA process, is being undertaken
at the time of writing, and will present a transport infrastructure network for
development in the medium term (up to 2015).

Funding Programmes

The majority of finance for development is to be found from the respective
national budgets supported by commercial loans and loans from international
financing institutions, including those from the World Bank and in particular
the EIB and EBRD, both of whom may impose less stringent economic analysis
criteria when considering applications for loans under certain conditions.

Encouragement is also given to the use of private equity contributions
through concession and partnership arrangements, although the take up of this
form of financing is becoming increasingly difficult. This is largely because
of continuing economic and political uncertainty, coupled with the failure of a
number of schemes in the region, generally as a result of low traffic projections
and the difficulties of governments to provide adequate guarantees.

The European Union however is making significant contributions through
a number of aid programmes. For the countries of pre-accession, through the
ISPA financing measures, up to 75 per cent of the cost of a project can be
met from European grant finance. In some of the poorer countries such as
Albania and FYROM, nearly 100 per cent of the cost is being met by grants
from PHARE programmes.

From the year 2000 onwards €1 billion will be available each year for
the following seven years for structural aid to the pre-accession countries
of the CEEC, with around 50 per cent of this being allocated to transport
infrastructure.

PHARE programmes will provide a further €1.5 billion each year. This aid
will be spread over the 13 countries of Central and Eastern Europe and will
include major contributions to the transport sector.

Project Implementation

Implementation is the responsibility of the respective countries along the Helsinki corridors and in the author's experience each country tends to run its own agenda regarding loan and aid coordination. As a result, infrastructure development has not necessarily been coordinated across borders and along specific corridors to date.

It is clear, however, that the TINA initiative has helped to focus for the first time on what is needed for future implementation across a large section of the network. Great emphasis is being placed on the inter-regional aspects for transport infrastructure development for TIRS. Countries are therefore increasingly being asked to look beyond their immediate national needs.

The implementation processes, however, do tend to be somewhat bureaucratic and Europe's aid programmes are no exception, involving Protocols and rigorous procedures to maximise transparency of the procurement process. Some improvements have been made in recent years with the introduction of a decentralised implementation system, which has taken many of the approval and administrative functions away from the bureaucracy in Brussels. That said, effective programme and project implementation then is dependent on the quality and experience of the local administrations, which in a number of CEEC countries, still need to be considerably strengthened in project management skills and in understanding international standards for construction contracts.

The Potential Network

The overall network of the Helsinki corridors extends beyond the network considered by TINA. Sections of some corridors fall within countries, which until quite recently have not qualified for European Union aid. The results of recent elections in the Balkans have changed those constraints. For example Branch B of corridor V and corridor X in Croatia and corridors VII and X in FRY have all been excluded from European Union funded investment programmes until now.

The potential network designated through the complete extent of the Helsinki corridor network now amounts to some 46,000 km of transport infrastructure, comprising some 19,800 km of roads, 22,500 km of rail and more than 4200 km of inland waterways. This is the result when the Croatia and FRY sections are added and when those corridor sections in Albania, Bosnia Herzegovina, and FYROM are included, all of which were excluded in the detailed analysis by TINA. These are the new sections being studied through TIRS.

Bottlenecks

Priority development of the corridors is aimed at eliminating bottlenecks identified through TINA and other more focused studies. Louis Berger's work on corridor V was one such study where difficult sections were found in many sections particularly in the north of Slovakia where the proportion of heavy vehicles in the traffic stream make journeys slow and quite hazardous with frequent fog and snow conditions prevailing for much of the winter months.

River crossings and rail over road bridges invariably are found to be typical bottlenecks where sub-standard carriageway widths provide an impediment to the free passage of two way traffic flows. Other hazards commonly found on these major arteries include slow moving cyclists and animal drawn vehicles, all of which compete for road space in heavily motorised traffic streams.

Removing these and other bottlenecks remains a major challenge. Similar difficulties can be found on the railways where inadequate signalling and historically poor track maintenance have left long sections of the rail network in need of essential improvement.

Construction Costs

The costs for developing these pan-European transport corridors are considerable. The costs identified through the TINA work (TINA Final Report, 1999) for the respective modes of road, rail and inland waterways for 10 CEEC countries (Bulgaria, the Czech Republic, Estonia, Hungary, Latvia, Lithuania, Poland, Romania, Slovak Republic and Slovenia) alone are shown in Table 3.2 below.

Table 3.2   Corridor development costs for TINA network (€ billion)

Infrastructure mode	€ billion
Road	43.5
Rail	37.0
Inland waterways	1.8
Total	82.3

It is has been suggested that these costs may be overstated. However, Louis Berger's experience throughout the six countries of corridor V and the preliminary work on TIRS reported in July 2001, suggests otherwise and when the development costs of the corridors in Southeastern Europe are included, it is unlikely that the total costs will be less than €90 billion, at present day prices.

Conclusions

Development Timeframe

TINA examined the affordability of making this scale of investment for each of the 10 CEEC countries of the Helsinki corridors by relating investment to a proportion of GDP in the respective countries. TINA assumed that the annual investment for development of the network should be limited to 1.5 per cent of GDP, with the target of completing development by the year 2015. Over an 18 year period between 1998 and 2015 TINA estimated the accumulated GDP on this basis. This was then compared with the estimated development costs for the Helsinki corridors for the pre-accession countries. This exercise demonstrated clearly that there are a number of short falls in the assumed available funding to meet these development costs in the specified time frame, including those in Bulgaria, Romania and Slovakia. Early results for TIRS suggest that the same problem will apply for most of the countries of Southeastern Europe.

For Bulgaria, Romania and Slovakia, the target of 2015 may be too ambitious for completing development of their sections of the Helsinki corridors. The same may apply to other countries such as Latvia and Lithuania and the countries of Southeastern Europe. Since the work of TINA was concluded, some 18 months ago, there have been further adjustments to many investment programs throughout the CEEC and Southeastern Europe, as the optimism for economic growth and development has fallen back from the heady days of the mid-1990s. Countries have become more cautious in a number of cases towards their attitudes to borrowing, with the result that investment priorities have been reviewed and programmes for development of transport infrastructure have been extended.

The consequences of these adjustments and other factors, particularly those affecting the corridor countries outside the pre-accession group, such as Albania, FYROM and Bosnia Herzegovina, mean that it may be some 20 years or more from now before development of the Helsinki corridors will be largely completed.

Risks

Despite these projections for development, a number of risks still prevail and may prevent infrastructure improvements from being achieved within these timeframes:

- *Political stability* must be paramount, particularly in Southeastern Europe; major infrastructure along the Helsinki corridor in Bosnia Herzegovina was extensively damaged during five years of war in the early 1990s. Significant elements still remain to be reconstructed, and this has resulted in major distortions to traffic patterns in that country and cross-border movements with adjoining neighbours. A recent report suggests that trying to rebuild the rail network in Bosnia will take so long that a whole generation will have grown up with no experience of rail travel, and indeed rail travel could simply cease to be viable in the region within the next 25 years.

    Elsewhere in the region the current difficulties in FYROM have resulted in suspension of construction activities along corridor VIII.
- *Economic growth* has to accompany political stability. Unless trade develops across the borders of Europe, the transit traffic along the corridors will not be generated and the justification for their development may not exist. Forecasting traffic growth in these regions remains one of the biggest problems when carrying out investment studies. Sensitivity analysis may provide a range of results but at the end of the day confidence in the results is essential, particularly when external funding is being sought. The continuing difficulties in the Balkans and the problems still being encountered in the Ukraine are key examples, which are hampering economic growth.
- *Conflicts of interest* between adjoining countries may result in delays to corridor development particularly in border areas. On corridor V for example, arguments still persist at a number of border crossings about the preferred location for road infrastructure to cross the border; this particularly applies when new alignments are being proposed. The environmental lobby may not want to see development which may become a particular problem for corridor VII, the River Danube, where necessary improvements to open sections to navigation would result in deepening or widening of the river.

Achievements and Future Prospects

Much development has occurred against a background of relatively low traffic flows, particularly away from large cities and industrial areas. Sections of the corridors have been developed as motorways, which could perhaps have been built as a single carriageway as an initial phase, thus releasing funding for much needed investment elsewhere on the corridors, where congestion would justify infrastructure upgrading. In some countries, previously committed construction has stalled due to lack of government funds to pay contractors. Unrealistic grandiose schemes, particularly some which promote private equity participation have been developed and in some cases tendered to the financial and construction markets, only to founder from lack of support.

These and other decisions may have been ill-considered and have resulted in inappropriate investments over the last decade, for road infrastructure development at least. The countries alone are not to blame. Funding institutions have been enthusiastic to see infrastructure development, sometimes disregarding or insufficiently testing the viability of investments.

However, development of Europe's transport corridors continues and the politicians' vision of a Europe united through an effective network of transport infrastructure is more than a dream. The EC's ISPA programmes throughout the 10 pre-accession countries of the CEEC are progressing, major investments in highway infrastructure in particular. In November 2000, the EIB reported that the initiatives following the Stability Pact for Southeastern Europe are progressing corridor developments in a number of countries in this deprived region. The current TIRS initiative is aimed at identifying further investment needs in the countries of Southeastern Europe to complement the earlier development needs identified by TINA elsewhere. Environmental issues are being sensitively handled, sometimes at great expense, across much of the region.

Enthusiasm abounds for the future. Among all of the 10 pre-accession countries of Central and Eastern Europe and the Balkan States of Southeastern Europe this can be almost infectious and is contributing to achieving development of the pan-European transport network across Europe within the next 10 to 15 years.

Notes

1   Under production by Apur, Paris France for the European Commission.

2   Bulgaria, Czech Republic, Estonia, Hungary, Latvia, Lithuania, Poland, Romania, Slovak
    Republic and Slovenia.

References

Berger, Louis, SA, TIRS reporting to date.

European Commission (1999a), 'Moving Forward. The Achievements of the European Common
    Transport Policy', EC.

European Commission (1999b), 'Traffic Forecast on the Ten Pan-European Transport Corridors
    of Helsinki', Multi-Country Transport Programme, NEA, The Netherlands.

European Commission (2000a), 'PHARE Multi-Country Transport Programme', *Newsletter*,
    January.

European Commission (2000b), 'Development of Branches of Corridor V', Multi-country
    Transport Programme, Prognos AG Consortium (including Louis Berger SA).

European Commission and Apur (2000), 'Semi Annual Progress Report on the Quick Start
    Package', November, EIB, Mapping (Slide 5), Paris.

European Commission Directorates General for Transport (1997), Towards a Common Pan-
    European Transport Policy, EC.

European Commission Directorates General for Transport and External Relations and TINA
    Secretariat (1999), 'TINA Final Report', October, Vienna.

Chapter 4

The Extension of Trans-European Networks to the Balkan Region: Planning and Financing Issues

Mateu Turró

Introduction

The European Investment Bank (EIB) is the policy bank of the European Union. Its main purpose is to foster European policies – in particular regional development, environmental improvement and establishment of better communication links – through long-term loans and other financial mechanisms adapted to the requirements of the projects. The EIB is a most important instrument of the Union to develop infrastructure of common interest. In the transport sector the approved loans during the 1995–2000 period reached almost €56 billion. EIB loans complement other financial sources, public or private, to facilitate the realisation of projects that are technically, economically and financially viable. The EIB, which is the first supranational financing institution in terms of lending, concentrates most of its activity in European Union countries, but is now confronted with the challenge of financing the major investment effort needed in European Union candidate countries to achieve the conditions for accession. It has also been requested to become a major player in conveying European Union assistance to the Balkan countries, to ensure peace and prosperity in the region.

EIB activity in Southeastern Europe – which is concentrated in Greece, the accession countries of the region, namely Bulgaria and Romania,[1] and Albania – has been extended to Macedonia, Bosnia-Herzegovina, Croatia and Yugoslavia. For the 10 accession countries and Albania, the EIB originally had a European Union lending mandate of €8.43 billion for the 2000–2006 period. For the accession countries and for 2000–2003, it also has an additional Pre-Accession Facility of €8.5 billion not guaranteed by the European Union budget. Macedonia, Bosnia-Herzegovina, Croatia and Yugoslavia have progressively been included in the European Union lending mandate, as part of the European Union Stabilisation and Association Agreements process (SAA) to help them in the reconstruction and upgrading of their infrastructure. As a consequence, the lending ceiling of the mandate has been raised to €9.28 billion. Loans for transport infrastructure should absorb a substantial share of this amount.[2] At the end of the period all countries should receive a share corresponding to their population and economy.

The EIB has also been requested by the Presidency of the European Union and the international community to lead the financial assistance programme of the Stability Pact, specifically in the transport infrastructure field. It is thus logical that the Bank, together with the European Commission and other IFIs, assists the countries of the region in a joint definition of transport infrastructure plans and accompanying investment programmes. The EIB contribution to policy definition and resource allocation is bound by the efficiency, sustainability and cohesion principles applied in the European Union. The practical experience of the Bank, in the appraisal and financing of transport projects all over the world, should be particularly useful in the Balkan region.

This chapter presents the situation of the joint planning effort being carried out in the Balkan region with the assistance of the EIB and the main criteria directing the exercise. The complexity and variability of the situation makes forecasting particularly difficult, but some estimates of future investment are also given, along with recommendations on how to speed up project implementation.

Planning the Transport Network in the Balkan Region

The need for improving transport infrastructure in the Balkan region is indisputable. It is also clear that the various countries in the region suffer from severe investment constraints. The relevant question is thus to define which is the most adequate strategy to deal with the substantial investment backlog within the available financial means. The international community and the European Union in particular are committed to the stability of the region and ready to provide assistance. This explains why, at the time, a list of 'quick start' projects EIB (2000a) was established in order to give an indication of the projects that seemed well conceived and could receive immediate financial assistance. This list included projects with clear economic or strategic interest and having the possibility of being implemented in the short term. Although lists of 'near-term' and 'medium-term' projects were also incorporated in the document, they were essentially an indication of potential projects that could be executed in a second phase. It was recognised, in particular, that medium-term projects should be the result of a more structured planning and programming exercise.

The EIB has thus launched, with the contribution of the European Commission and some donors, two planning exercises, one for air transport infrastructure (A-TIRS)[3] and another, more general, for the definition of transport investment priorities in the other modes: the Transport Infrastructure Regional Study (TIRS). The World Bank, on the other hand, is carrying out a 'Transport and Trade Facilitation Programme' that should also contribute to improve the performance of the transport system.

The first phase of TIRS is presently under way. Financed by the Agence Française de Développement and also supervised by the European Conference of Ministers of Transport, the Commission and the EIB, the consulting firm Louis Berger is responsible for its realisation. The first proposals of the study were presented in a regional conference in Bucharest on 12 July 2001.

A main objective of the study is to establish, from a multimodal perspective and taking environmental considerations into account, the basic inter-regional transport infrastructure networks that the Balkan region needs, in line with the TINA exercise. For this it is necessary to identify major international and regional routes in the region and define a coherent medium term network to be used as a framework for planning, programming and coordinating infrastructure investments.

The need for a global view on the problems of transport is clearer in the Balkan region than anywhere else in Europe. The small size of most countries, the complexity of the topography and the need to overcome past conflict between the various states all plead for a specific planning method. This method is bound to combine a top-down concept of the transport system, aiming at global performance without the artificial constraints of political borders, with the bottom-up considerations and requests from the various countries and regions. The TIRS study, which focuses on inter-regional links, has thus produced a broad study network on which information has been collected and from which a backbone network, similar to the TINA network for the accession countries will be defined.

A second objective of the TIRS, given the importance of immediate action in the region, is the definition of short-term priority projects suitable for international financing. Projects located in the backbone network (or, under certain circumstances, giving direct access to the network), representing immediate and urgent investment measures, necessary to bring the existing major long-distance routes to an acceptable level, will be considered as priority projects for European Union financing. In this first phase, the TIRS will mostly select rehabilitation projects, which are often those with the highest profitability when the physical condition of the infrastructure is poor. The selection will take into account a standard cost-benefit analysis and other considerations, in particular the need to ensure minimum quality conditions to the backbone network. The selected schemes will then be subject to a feasibility study to confirm its interest and address in detail the technical, environmental, economic and financial aspects of the project.

The study will thus require a second phase that will look into the longer term and incorporate some of the upgrading and development projects expected by the countries of the region requiring funds that are presently unavailable. The study includes the definition of this second phase, which should start early next year and be financed by the European Commission.

The European Commission, which is actively participating in the TIRS, has contributed to the planning exercise with a document 'Transport and Energy Infrastructure in Southeastern Europe', providing a global strategy for the networks in the region, based primarily on political, geographical, demographic and regional (that is, socioeconomic) considerations. The recommendations in the document are in line with those guiding the TIRS, but the document proposes a basic network using both technical and political criteria. This 'strategic' network is obviously a small part of the study network considered in the TIRS, which includes most links that are relevant for long-distance traffic. The 'strategic' transport network based on a top-down approach will thus have to be confronted with the results of the analysis of the study network in TIRS, taking into account the actual demand and the practical constraints, to see if it should be slightly modified to better respond to these considerations.

Indeed, the success of the ongoing planning depends heavily on the ability of the countries of the region and the international institutions to jointly establish a process for the efficient application of resources, with the specific aim of developing the relations between the Balkan countries and the European Union.

The Planning Context

For the time being this process will be mostly directed at infrastructure improvement. For the next few years, the investment potential in the region will be rather limited and, in the case of transport, it has to concentrate on the rehabilitation of its infrastructure assets and the upgrading of selected superstructure and rolling stock. It is, however, clear that the final aim of any transport policy is to provide good transport services; adequate infrastructure and vehicles is but a small component of the requirements. The application of adequate regulatory and pricing policies is often more effective, at least in the short term, than heavy investment. This requires the creation of a new, adequate and more efficient regulatory, organisational and institutional framework that would facilitate the implementation of reforms towards market mechanisms and the development of modern practices in the transport sector, including measures to preserve the environment. This adapted framework is also essential to ensure the efficient implementation, management and maintenance of transport infrastructure.

A particular emphasis should be given to the solution of border crossings. This is particularly necessary for the region, since many new borders have been created after the dissolution of the FRY. It is obvious that the facilitation of trade and traffic flows brought about by infrastructure improvements will be useless if border crossings continue to act as bottlenecks. All efforts have to be made to reduce waiting time at the borders by institutional changes and the use of best practices and modern technology.

The European Union and the IFIs insist on these aspects and demand the adaptation of the administrative and legal context, with the application of performance-oriented criteria in the public sector, and the introduction of competition wherever possible. In the case of infrastructure, past experience in the region has shown the difficulties of project implementation. This explains that the donors insist on having a strong input that is considered indispensable to ensure reasonable chances of successful project completion. If the public sector continues its transformation and is able to adequately manage project preparation and implementation, the interference will certainly all but disappear. On the other hand, a persistent lack of capacity of the administrations to efficiently manage projects could become a major restriction to European Union assistance.

Criteria for Project Selection

The Steering Committee of the TIRS is presently dealing with the difficult subject of establishing criteria for the selection of priority projects. It seems clear that a combination of conventional criteria, such as cost/benefit analysis and more political ones, mostly directed to regional integration, should be applied. These are some basic considerations that support the discussion:

- *in the short and medium term*: the first phase of the TIRS study will produce a backbone multimodal network. Proposed rehabilitation and renewal schemes on this network will be analysed through pre-feasibility studies. The selection of those with the highest priority for earliest implementation should take into account a standard cost-benefit analysis[4] but also other criteria, in particular the need to ensure minimum quality conditions to the backbone network. The selected schemes will then be subject to a feasibility study to confirm its interest and address in detail the technical, environmental, economic and financial aspects of the project;
- *in the long term*: the second phase of the TIRS study will be a full regional planning exercise, coordinated by the European Union, in which the future transport infrastructure requirements, linked with expected economic growth and integration with the European Union, will be addressed. It will contemplate the possible expansion of the basic multimodal network, the definition of a secondary network and the upgrading or new construction of some components of the network. An investment programming exercise based on more systematic pre-feasibility studies and taking into account the network effects, will define the future investment context for the various countries and the expected financing sources. Feasibility studies will also have to be produced to justify each individual project. The technical standards and the quality of transport infrastructure assets should correspond to the expected traffic and ensure adequate socio-economic rates of return to prevent a misallocation of scarce economic resources. Feasibility studies must also ensure interoperability conditions in all modes: railways (electrification, signalling, etc.); roads (axle loads, signing); inland waterways (clearance, draught) and aviation (ATC systems).

It is important to stress here that feasibility studies are only useful when based on reliable traffic forecasts. The proper recollection of transport data and the updating of traffic forecasts are fundamental for any medium- and long-term planning exercise. The creation of a permanent institutional framework for the observation of transport developments in the region appears to be a suitable solution for this objective.

Financing Issues

The TIRS will carry out a screening process of the projects, mostly located in the backbone network, to determine if their designs are adapted to the needs and to establish their economic viability.[5] The number of projects will be limited by the financial envelope each country can devote to inter-regional transport infrastructure investment (around 1.5 per cent of GDP per year, if TINA estimates are followed). The priority projects, selected after the screening process and the pre-feasibility analyses, will finally be integrated into an investment programme for the whole region for the next few years. This programme will be proposed for discussion at the final TIRS conference in early 2002, and bilaterally between the countries and the international institutions providing financing.

It must be stressed that the selection of a project as a priority in the TIRS does not necessarily mean that it should be carried out immediately or that it can count on financing. A proper feasibility study to produce a reliable estimate of costs, a socioeconomic evaluation and a finance structure that guarantees the completion of the project confirming the initial estimates will be needed.[6] This should not be heavy work, but it must be able to answer the key questions that the financial institutions usually ask of promoters. To facilitate the work of the evaluators, it is recommended to use the TINA guidelines for socioeconomic cost-benefit analysis.[7]

In any case, the macroeconomic environment is extremely fragile, which underlines the desirability of grants or very soft loans, at least as sizeable cofinancing. It also means that the capital base of infrastructure should not be widened beyond the means of the economy to support the corresponding annual maintenance cost streams, which can be quite considerable and amount to several points of GDP.

To contribute to the improvement of the decision making process, the EIB has recently launched the FAST (Financial Assistance for Preparatory Studies) mechanism. It will provide up to €2 million/year to finance preparatory studies (technical, economic and financial) directly linked to projects specifically in accession countries and the Balkans. This up-front financing will be reimbursed if and when the project is implemented. In the Balkan region, in particular when the promoter has substantially contributed to the financing of the studies, the reimbursement might not be requested.

Conclusion

The EIB is one of the main vehicles of the European Union to provide assistance to the Balkan region, in particular to the improvement of its transport infrastructure. The main interest of the EIB is in financing projects that are technically sound, economically viable and which have a solid and sustainable financial structure. Under the present circumstances, the need to catch up with the backlog in the region requires the concentration of the limited resources available for transport infrastructure in rehabilitation and small and selective upgrading where justified. An adequate selection of projects is, in any case, essential for optimisation of efficiency. The EIB, along with the Commission, the ECMT and other institutions, is supporting a planning exercise, in two phases, that should assist the Western Balkan countries to establish a coherent multimodal transport network supporting the internal economic and social integration of the region and facilitating its wider integration with the European Union.

Notes

1   Slovenia is not included in most groupings referring to the Balkan region.
2   In 2000, transport absorbed 48 per cent of the total.
3   The 'Air Traffic Infrastructure Regional Study for South- Eastern Europe' was produced by Nordic Aviation Resources and finalised in May 2001.
4   Using the TINA guidelines 'Socio-economic cost-benefit analysis', October 1999.
5   Being mostly rehabilitation projects the environmental component is less relevant.
6   The types of projects envisaged do not allow, in general, innovative financial formulae. Besides possible grants, most resources will have to come from public budgets or publicly guaranteed loans.
7   A specific adaptation of the guidelines for rail projects, supported by the Commission, the UIC and the EIB, was produced during 2002.

References

Croatian Academy of Sciences and Arts (2000), 'Traffic Connection between the Baltic and the Adriatic/Mediterranean', Zagreb, November.
Cvetanovic, O. (1999), 'Consequences of War Destruction for the Yugoslav and Regional Transport Infrastructure', in Minic, J. (ed.), *Southeastern Europe 2000. A View from Serbia*, European Movement, Belgrade, .
EIB (1999), 'Long-term Development Issues for Southeastern Europe', July.
EIB (2000a), 'Basic Infrastructure Investments in Southeastern Europe', February.

EIB (2000b), 'Western Balkans Transport Infrastructure Inventory', July.

EIB (2001), 'Air Traffic Infrastructure Regional Study (A-TIRS) for Southeastern Europe', May.

European Commission (2001), 'Transport and Energy Infrastructure in Southeastern Europe', October.

Group of Four Infrastructure Experts (1999), 'Infrastructure Atlas', Autumn.

TINA Guidelines (1999), 'Socio-economic Cost-benefit Analysis', October.

Turró, M. (1999), *Going Trans-European. Planning and Financing Transport Networks for Europe*, Pergamon, Elsevier.

Chapter 5

The Czech Transport Infrastructure and Pan-European Corridors

Antonín Peltrám

Introduction

The transport system of the Czech Republic inherited an infrastructure which is oriented rather in favour of railways, especially heavy goods trains, with an ability to cater for bulky products and, in where problems arise, with remarkable preference for quantity to quality (heavy trains, low speed, the quality of track adjusted to accommodate lower velocities, naturally within the UIC (International Union of Railways) scale). The infrastructure in all modes of transport has been poorly maintained. The lack of money for constructing motorways, for example, more quickly was to some degree compensated by many constraints on the development of all types of road transport. After the 'Velvet Revolution' these constraints have broken down; the gap between the capacity of and demand for infrastructure has opened up at exactly the same time that many more actual problems connected with the transformation from central command to the free market economy have occurred. It is now necessary to find money for maintenance, improvement and new construction of main railway lines with a higher quality of services offered both in passenger and in goods traffic and roads and motorways – all in an environment with many more constraints than at any other time.

A substantial part of the growth in traffic could be derived from the geographical position of the Czech Republic in the growing transport needs of unified Europe, with some compensation for additional costs from foreign users of infrastructure. To speed up the adaptation of the most extensively used parts of the infrastructure – those connected with international traffic – some subsidies from the EC could be used. One of the anticipated preconditions for such assistance is to accept *acquis* in transport. Therefore, the first efforts of the present government were directed towards full harmonisation of Czech transport law with the EC law. This has been achieved: 14 amendments of recent acts in transport, telecommunications and postal services have been accepted. The Czech Republic has not applied for any exemptions or temporary measures in transport. Therefore it has not been necessary in this chapter to draw attention to the harmonisation of laws, as it had been intended, with the exception of the as-yet-unsuccessful restructuring of the Czech railways, but probably successful attempt to assure a permanent flow of money for financing transport infrastructure in the form of state fund of transport infrastructure.

As to the infrastructure, there are two main problem areas: the first concerns the trans-European network passing the Czech Republic and the second concerns the regional and local network – for the pan-European level, the secondary network. But such a network is necessary both for commuting and as a feeder for main railway lines. Because of large hidden loans created by the undermaintenance of transport infrastructure, it has been necessary to create a basis for funding which is in some way independent of the yearly approved and limited state budget. Such a basis has been created by an Act establishing the state funding of the transport infrastructure. Taking advantage of this intiative, the Czech Republic itself should contribute the necessary amounts – both single lump sum subsidies from the sale of state shares in some companies and stabilised income, as a share of taxes and duties levied in transport. Such resources could be combined with cofinancing from ISPA (Instrument for Structural Policies for Pre-accession) funds and supported by the public-private partnership.

The Czech railways are heavily subsidised every year and, despite such assistance from the state budget, suffer heavy losses. Therefore there are shifts of money from maintenance, improvement and investment to personal costs. This threatens the future of railways, especially regional and local lines. Some measures should be accepted in connection with the limitation of the deficit of the state budget for the next year. A solution could be found in a medium-term financial plan, with blocked expenses in the medium-term state budget. A temporary solution could be found in state funding of transport infrastructure, with clear proportions for financing railways and roads. (The advantage of this fund is in the possibility for taking loans on account of future fixed income.)

Recent Developments

During the period of the central command economy in the former Czechoslovakia, there had been a very steep increase in the volume of goods traffic in both main modes of transport: railways and roads. Inland waterways were normally used for connections with seaports in Germany or Poland and used the network of channels in Germany, Netherlands and other West European countries to reach other seaports. Somewhat temporary inland operations were connected with the combined traffic rail/waterway of brown coal (because of lack of capacity on the main railway corridors parallel to the navigable stretch of rivers, with only very expensive possibilities to improve the situation by the construction of third and further tracks) and sand and gravel extracted from the pits in the direct surroundings of the navigable stretches of the river Vltava (Moldau) and Labe (Elbe). Raw and energy-intensive types of the development of the national economy was reflected in the structure of goods transported. In goods traffic there was a predominant, steadily growing share of bulky products: coal, ore, construction materials, and had followed a considerable difference between a very high gross product in comparison with a rather low net product; there are many other specific symptoms of transport-intensive types of development. Development was in favour of railways (an inland state with only some 100 km of navigable rivers has few other possibilities). Because of the lack of ore deposits (deposits of suitable quality for commercially-efficient mining having long been exhausted) and other raw materials and energy, with the exception of coal, there was a considerable amount of imports of raw materials from the previous Soviet Union and exports of heavy products to the COMECON countries, and railways were preferred to other modes of transport. It is possible to suppose that the development of the transport system in favour of the railways was taking place even under the influence of the military doctrine of Warsaw Treaty. Despite upgrading the railways in preference to other types of transport system the capacity did not grow in such a way as to eliminate delays and other symptoms of bad transport quality.

But there had been some attempts to regulate traffic volumes, to reduce the demand for transport – something that has actually appeared within the proposals of the European transport policy.

Lack of Investment Capital

Use of central state investment funds in favour of heavy industry caused a lack of finance for the transport system in general, and roads and highways in particular. But because a halt was called to the development of road transport, with a system of assignment of capacities of transport and allocation of the purchase of all types of road vehicles controlled by tough planning, plus some attempts to regulate demand across all modes of transport, the gap between supply and demand of capacities has never been so wide as at present.

Preference of Quantity over Quality

As has been mentioned above with regard to the structure of traffic and lack of capacities, there were preferences in favour of railways and artificial constraints on developing road transport. There was support for inland waterway transport, but the navigable stretches of the both above-mentioned rivers, Labe (Elbe) and Vltava (Moldau), with about 303 km (without lakes or dams) in the Czech Republic and river Danube in Slovakia (172 km – mostly as a natural border with Austria and Hungary) are too short for massive inland domestic traffic; therefore the main orientation of the inland waterways traffic is in favour of export and import services.

Development of Railways

The state railways were upgraded with special regard to quantity: electrified main double track lines were used predominantly by heavy goods trains, with great share of block trains, at a time when the capacity of lorries was limited (production of the heaviest class was oriented towards export; there were limited import possibilities from the COMECON countries; and in fact no possibilities for import from the states outside COMECON). But even in the production of lorries orientation was towards quantityrather than quality. All these features changed after the 'Velvet Revolution'.

The 'Velvet Revolution' and Changes in Transport

As to passenger transport, for political reasons that had influenced constraints on the construction of necessary capacities and the limited import of cars, there was a very high share of collective types of transport: railways, buses, urban mass transport. It copied in some way the high degree of concentration of employment in large factories. But, nevertheless, such constraints on the development of road transport were against the citizens' wishes to own a car – mostly as a symbol of freedom of movement; for a long time using cars had been taken as a type of luxury consumption, imported cars were only for people able to pay in foreign currency, or on special allowances.

Full Harmonisation of the Czech Transport Law with Community Law

Important change in Czech transport policy could be based only on the substantial change of transport law oriented towards full harmonisation with Community law. It is necessary not only as a precondition for the accession of the Czech Republic, but because a transport system of such a small state in the heart of Europe could not properly operate without acceptance of the principles of free market economy, naturally with some necessary modification for transport.

This target was reached by amending previous or accepting 14 new Acts dealing with transport, communications and postal services. It would be necessary to assure only relatively small changes according to the development of *acquis*. What was not successful was the restructuring of the Czech railways (CD). It is ironic that, the later the transformation, the worse for the future of the Czech railways in general. But it is a problem that must be solved soon: as noted in the 2001 Regular Report on the Czech Republic's progress toward accession:

> The legislative delays underlined in the previous Regular report have not been redressed. Furthermore new *acquis* has been adopted by the European Union in 2001 concerning the establishment of a liberal market. Thus, substantial legislative work remains necessary in order to achieve alignment ahead of accession. In particular further steps need to be taken to implement the new railways *acquis* in order to ensure independent functions of the infrastructure manager as regards allocating of capacity and charging (EC, 2001b).

On the other hand, the Regular Report admitted the function of a target-oriented income for financing the transport infrastructure; in the Czech Republic it has the form of the state fund for transport infrastructure described below – in a better way than has been accepted by some international financial institutions. (The financiers recognise loans and debts in a form of financial obligations rather than in a form of hidden loans.) And a good support for such fund was expressed in the European Commission's White Paper 'European Transport Policy for 2010: Time to Decide' (EC, 2001a). Therefore, I shall try to describe this legal tool of the Czech transport policy.

Transformation of the National Economy

During the process of the transformation from a central command to a free market economy after 1989, there was a substantial drop in the structure of production and directions of foreign trade, and a decrease in the production of the largest companies supplying the previous COMECON markets, etc. The division of the unique Czechoslovak Republic into two states in rail traffic has brought a large decrease in the average length of goods traffic and has supported, in fact, relatively unlimited development of all types of road transport. All the elements changing the economy and the extent and structure of employment reflected in the transport sector are in some way expressed in the indicators in the next tables.

Because of the decision to divide the previous Czechoslovak Federal Republic from 1 January 1993, the time series from the year 1990 are difficult to reconstruct – simple addition of data for both states is not really adequate. Therefore it looks preferable to introduce the time series from 1 January 1993, date of the origin of the Czech Republic and the Slovak Republic, for the Czech national economy and transport.

Table 5.1   Czech railways' performance development (billion passenger – km and billion tonne-km)

Year	1993	1994	1995	1996	1997	1998	1999
Passenger km	8.55	8.48	8.02	8.11	7.72	7.02	6.96
Tonne-km	25.14	22.70	22.63	22.34	21.01	18.76	16.71

Table 5.2   Road and urban mass transport (thousand million passenger km)

Year	1993	1994	1995	1996	1997	1998	1999
Buses	9.09	8.20	7.67	6.32	5.88	5.98	5.95
Urban mass transport	2.64	2.58	2.26	2.21	2.16	2.18	2.26
Cars and motorcycles	49.00	51.70	54.50	57.90	59.00	60.80	62.30
Total	60.73	62.48	64.43	66.43	67.04	68.96	70.51

The rate of increase of road transport is apparent from the content of Table 5.3.

Table 5.3   Registered numbers of vehicles (excluding approx. 1.5 million motorcycles)

Year	1993	1994	1995	1996
Cars + vans	2,746,995	2,923,916	3,043,316	3,192,532
Lorries	147,657	184,278	202,929	225,477

Year	1997	1998	1999	2000
Cars + vans	3,391,541	3,492,961	3,439,745	372,316
Lorries	246,621	260,276	268,259	323,255

Infrastructure Capacity

Limited increase in numbers of road vehicles in past decades facilitated the low rate of increase of capacity. Transport infrastructure is a very convenient way to balance the public budgets: according to the estimates based on the expertise of the World Bank – inner debt, hidden loans – the decrease in infrastructure maintenance in 1990 was about 350 billion Czech Crowns – nearly €10 billion. This inner debt from the 'Velvet Revolution' has risen. There is not enough money for maintenance of railway infrastructure, nor for improvement and new construction of railway infrastructure. (The situation is similar all over the world – in general, no states have enough money for investments, and on the other hand, in most states which previously operated under a central command economy, because of past 'savings' money for maintenance, there are in general more or less huge delays in funding the infrastructure and it is necessary to fill these gaps. The Czech Republic's present situation is different only in its intensity.) Nevertheless, such a steep increase in demand for capacity in road infrastructure as has occurred in the Czech Republic has been seen very seldom. And it is a huge task for the Czech Republic, if we compare the extent of infrastructure and the relevant items in the state budget, to cover the costs of maintenance, improvement and development. In Tables 5.4 and 5.5 the present length and structure of transport infrastructure by modes of transport and amounts spent from the state budget is described.

Table 5.4   Transport infrastructure network in the Czech Republic in 1999

Railways
Length of operated lines	9,444 km
Standard gauge	9,342 km
Single track	7,515 km
Double and more than double tracks	1,929 km
Narrow gauge	54 km
Total length of constructed tracks	17,025 km
Total length of electrified lines	2,843 km
in which double tracks and more	1,713 km

Roads and motorways
Total length	55,432 km
Motorways	510 km
Roads (without local roads)	54,933 km
1st class roads	6,005 km
2nd class roads	14,686 km
3rd class roads	34,242 km
Local roads	72,300 km

Navigable inland waterways	663.6 km
Airports	84
of which for scheduled international flights	4
Combined transport terminals	14

Development of Transport Networks

According to the high density of the infrastructure network (the state area is about 78,876 km²), but also because of limited amounts for unrestructured railways the total length changed only by the construction of about 200 km of new motorways and highways, diminishing the number of terminals of combined transport and some tens of kilometres of railway lines.

The improvement of railways is oriented towards the two main corridors – parts of the Helsinki corridors IV and VI. In some years – with a few years' delay – they will be fully upgraded to allow speeds of up to 160 km per hour. Some stretches could enable a maximum speed of 200 km per hour, but because of the high density of railway stations on the lines and short distances between them, it would not be of particularly useful to exploit these possibilities for regular traffic.

Table 5.5 Investment support from the state budget (million Czech Crowns; €1 = 36.5 Czech Crowns)

Year	1994	1995	1996	1997	1998	1999	2000
Czech railways	2,392,100	3,221,200	3,510,100	4,318,000	4,503,300	5,680,423	5,695,500
Roads and motorways	6,491,956	8,957,237	9,595,732	10,689,548	6,434,013	9,553,632	7,405,607
Airports	365,000	297,599	387,300	314,521	125,375	93,114	237,900
Waterways				87,384	170,864	115,000	205,305
Combined transport terminals			5,000		17,958	38,000	38,000
Metro							300,000

Table 5.6 Czech Railways – subsidies in operation (excluding investments) in 1,000 Czech Crowns

Year	1994	1995	1996	1997	1998	1999	2000
Total subsidies	5,771,000	5,288,250	5,218,529	5,652,014	6,430,296	7,086,430	7,910,998
Research and development	6,671	5,300	7,220	5,510	4,641	6,480	
Interest, debt service				200,000	300,000	1,218,986	2,936,780
Total	5,771,000	5,294,921	5,223,829	5,859,234	6,735,806	8,310,057	10,854,258

The main motorways following, in principle, the path of both above-mentioned corridors (Nuremberg – Bavaria borders – Rozvadov – Prague – Brno – Bratislava and Saxony – Děčín – Prague) are under operation, with the exception of Pilsen bypass and Czech Central Mountains throughpass because of the late conclusion of many years of discussions and trials with the NGO. Both missing parts are now under construction.

The possibilities to finance the construction of stretches of trans-European corridors using money from the state budget and cofinancing from ISPA funds could shorten the period of reconstruction. That is very important from the point of view of the Czech national economy development and transport needs of unifying Europe.

Possibilities of financing only by means of public expenditure are partially limited because of the huge subsidies and deficits of the Czech Railways (CD); a part of them is covered on account of the state property operated by this state organisation.

Permanent Incomes for Infrastructure Improvement and Development

Financing the transport infrastructure either from the state budget or with its participation has had some problems: the state budget is set out for one year only; the Czech Republic has had no financial perspective plan (that is, for more than one year).

Discussion of what the breakdown of any admissible state budget deficit might be cannot be achieved sooner than spring – the second financial quarter. This makes preparation of reconstruction and construction works for the year in discussion rather late, particularly as the budget must be spent by the end of the year. This timing can be be too short for rational disposal of approved amounts.

Such a limitation could be eliminated using the state fund or another arrangement with money not tied to the calendar year. Moreover, because of assured incomes it would be possible to take loans not connected with the state guarantees and the reference values of the excessive deficits of government budget and debts.

To eliminate the these problems the Czech parliament has passed an Act of state funding for transport infrastructure.

The State Fund for Transport Infrastructure

This fund comprises amounts that were formerly, or have been, intended as the incomes of the relevant chapter of the state budget. It is a form that must be filled with appropriate content – necessary amounts of money. It could be very useful at least until it is possible to prepare medium-range budget planning with the amounts which remain after the end of the financial year in question.

The state fund for transport infrastructure should be made up of:

1   short and therefore temporary incomes. They should come from the sale of present state property shares in those remaining undertakings which retain a share of state ownership, mainly in the communications sector (for instance, radio communications and telecoms). The government promised them on the level of 30–40 billion Czech Crowns. But there have been some other gaps in the state finances and the conditions for property transactions in the telecommunications sector have not been the most favourable, because of massive recent investment;
2   permanent incomes, comprising:
    – a share from the excise duty on mineral oil products;
    – charges for using motorways and express roads;
    – road taxes (taxation of all vehicles).

The main item represents the share of excise duty on mineral oils products. For 2002, the amount was proposed by the Ministry of Finance and later approved by the government and parliament at a level of 20 per cent of the total income of the excise duty. It is too low – the estimated volume should be at least 27 to 30 per cent; but the under-financing can only be corrected in the future – hopefully in the near future. In the meantime the uncovered amount could be compensated for only by an earlier transfer of money from the sale of state shares in some enterprises with state property participation, as has been mentioned above. But it depends on the early sale of these shares and on receiving a good price.

The fund could take loans on account of future income assured by law in favour of constructing works financed by means of credit; but the precondition for such operations should be a fixed satisfactory future income.

The budget of the state fund for transport infrastructure is approved by the steering committee with the members from main parliamentary parties in the Czech parliament, and then by the parliament at the same time as it approves the state budget, with the exception that the planned expenses can be carried over to the next year.

We also anticipate cofinancing from ISPA or PHARE or other forms of assistance from the European Union, but it could be only a part of necessary funding in form of supplementary financing and mainly because of the close connection of such cofinancing with pan-European interests. And for the most urgent project we hope that the cofinancing would be implemented much sooner than in the previous year – in the reality, the first year of ISPA support.

We are glad to see that the European Commission White Paper 'European Transport Policy for 2010: Time to Decide' supported the idea of dedicated budget incomes.

Necessary Transformation of the Czech Railways

Until the transformation of the Czech railways into infrastructure administration (accounting fees for using infrastructure) and business-oriented organisation for railway traffic operations which allows competition with other railway traffic operators (a problem that should be solved at least in connection with the preparation of the accession of the Czech Republic to the European Union) it is difficult to apply any form of public-private partnership, with the exception of loans from private banks, but guaranteed by the state – and therefore limited by law.

Analogous to the railway network there is a scheme of first and second class motorways and roads; the difference illustrates the part of network that could be accepted as a network of common interests of the Czech Republic and European Union. Only the Czech Republic's own resources could finance the predominant part of the network that is of non-European importance (see the characteristics of the network in Table 5.4).

However, as in all the neighbouring states, and having regard to the Community law, it is necessary to raise the taxes and charges on road transport, especially for selected infrastructure and heavy vehicles. The higher the charges, the more important it is to introduce charges for only a limited period of a year (daily, weekly, monthly charges). And to control the payments for shorter period of time than the whole year in particular, and without distortions of traffic flows, it is necessary to use electronic fee collection. We anticipate the introduction of a system compatible with the German system; probably very similar to the future Austrian system and compatible with the existing

Swiss system. We hope that the highest level of compatibility will be reached with the future first stage of GALILEO.

There is another specific problem: because of the frequently stressed problem of the overused capacity of Czech road transport, it could be necessary to limit the increase of capacity in road transport, as was already set out in the Treaty of Rome and later on enabled by European Council Regulation No. 3916/1999, otherwise there are possible conflicts with participants in road transport.

Using the state fund for transport infrastructure with an appropriate level of incomes, as has been suggested, will enable the full use of the public-private partnership.

The capital investments in improvements in transport infrastructure naturally are not the only conditions for a cost-effective system in transport. But they are one of the most important preconditions to reach such an ambitious target.

Conclusions

The density of the Czech transport infrastructure is very high. But in the past decades there was no money to cover necessary maintenance; the cuts in budget for transport were predominantly used mostly in favour of heavy industry. Now it is necessary to find permanent incomes as a first step to stop further worsening of this network, and later on to decrease the previous delays. As to the railways, they should be connected with some kind of selection of the secondary network which could survive, with the transformation of the Czech Railways into business-oriented organisation and the stabilisation of the necessary funds for constructing and reconstructing the network that should be retained. It is clear what should be retained from the point of view of TEN and there should be – maybe with some delays – not insurmountable constraints for joint sources for financing it. But as elsewhere, the motorways especially will attract more traffic, as could be maintained on four to six lanes motorways. It is necessary to join the pan-European effort for greater share of railways. The decisive parts of TEN railway corridors should be reconstructed within some years. Financing of the other parts is a problem. There is a strong initiative to create a way to provide uninterrupted financing of the infrastructure during a period of several years in the newly-established state fund for transport infrastructure. Naturally it must have appropriate content – in this case necessary level of incomes. We should try to achieve them.

References

Czech Republic (1999), 'National Development Plan of the Czech Republic – Transport, Telecommunications, Postal Services', September.

European Commission (1998a), 'Socially-necessary Railways in Europe', SONERAIL, Project of the 4th Programme of Research and Development of the EC.

European Commission (1998b), 'Pan-European Transport Policy. Prospects and Priorities for East-West Cooperation', December.

European Commission (2001a), 'European Transport Policy for 2010: Time to Decide', European Commission White Paper, September.

European Commission (2001b), 'Regular Report on the Czech Republic's Progress towards Accession', November.

European Commission (2001c), 'The Future Development of the Common Transport Policy. A Global Approach to the Construction of a Community Framework for Sustainable Development', European Commission White Paper.

Peltrám, A. (1999a), 'Koncepce rozvoje dopravy a spojù (etapa roku 1999)', ['Concept of the Development of Transport and Communications (Stage of 1999)'], *Veřejná správa* [*Public Administration*], Vol. 19, No. 8, p. 3.

Peltrám, A. (1999b), 'Koncepce rozvoje dopravy a spojù (etapa roku 1999)', ['Concept of the Development of Transport and Communications (Stage of 1999)'], *Veřejná správa* [*Public Administration*], Vol. 10, No. 8, pp. 10–11.

Peltrám, A. (2000), 'Discussion Document regarding the Process of Restructuring Czech Railways', January, Prague.

Chapter 6

Prospects for the Development of the Southeast Section of Rail Corridor IX

Christos Pyrgidis and George Giannopoulos

Introduction

This chapter presents some of the results of a study project[1] concerning the (pre-)feasibility of the development of the railway corridor no. IX in its section between the Greek port of Alexandroupolis in the north, and Moscow (Russia) (AUTh, 2001). The route of this corridor is: Alexandroupolis – Bucharest – Chisinaou – Kiev – Moscow, including a branch to Odessa. The aim of the work was to analyse the rail transport characteristics for passengers and goods along this 'southeast' section of corridor no. IX.

Another aim was to investigate the problems and possibilities that exist for the improvement of this section, and its functioning as a credible alternative to road transport for the movement of passengers and cargo. Particular attention was given to evaluating the exploitation and other development characteristics of this corridor as a land connection between Greece and the other Southeastern European countries.

Data were collected from the various railway administrations along the corridor, and via independent channels and individuals.[2]

The rail networks along the ten Crete corridors, with the support and participation of the UIC (International Union of Railways), the OSJD (Cooperation of Railways Organisation) and other interested international organisations have formed special committees for the promotion of actions and cooperation that will enhance the proper functioning of these axes as international rail corridors. The committee for corridor no. IX is chaired by the Greek Railways.

All countries and rail networks situated along the corridor have prepared a series of improvement programmes referring to both infrastructure and rolling stock. However, priorities are in general not always the same and coordination of these plans is far from effective. Thus, the successful promotion of the corridor towards acceptable performance standards that would make them able to compete with road and maritime transport or (even) air, is a matter of special study and attention and it was the focus of the research study mentioned above.

The development of corridor no. IX into a credible transportation route for international freight and, to some extent, passenger transport is thus primarily an exercise in international cooperation and the ability of the 'partner' rail networks to adjust to the challenging requirements of modern railway transport. There are considerable funding needs that require complex financial engineering and negotiations. However, examples of other successful similar efforts exist and it is believed that Southeastern European railways can profit from these experiences. In its recommendations the study has taken into account these experiences, the primary lesson of which is the need for efficient cooperation between all sides.

A Critical Assessment of the Current Situation

General Description of the Corridor

Although the length of the section considered in the study is approximately 2,900km, the total length of rail corridor no. IX is 6,350km. It is the largest European rail corridor in terms of length (UIC, 1994). It crosses nine countries (Greece, Bulgaria, Romania, Moldavia, Ukraine, Lithuania, Belarus, Russia and Finland), and an equivalent number of border crossings, with twice as many border stations. It also operates in two major gauge widths, something that makes its uniform operation from end to end difficult and problematic.

The total length of the corridor can be divided into discrete sections (see Figure 6.1).

- *southeast section (Alexandroupolis – Moscow)* has a total length of 2,920km and consists of a main section of 2,856km (Alexandroupolis – Roldilna – Moscow) and a small branch of 64km that links the main line to the port of Odessa (Roldilna – Odessa):
- *north section (Helsinki – Moscow)* runs from Moscow to Gologoye to St Petersburg to Helsinki and has a total length of 1,092km;
- *west section* has a total length of 1,056km and consists of a main section with two branches, as follows:
  - main section: Kaisiadorys – Zlobin;
  - north branch: Klaipeda – Kaisiadorys;
  - south branch: Kaliningrad – Kaisiadorys;

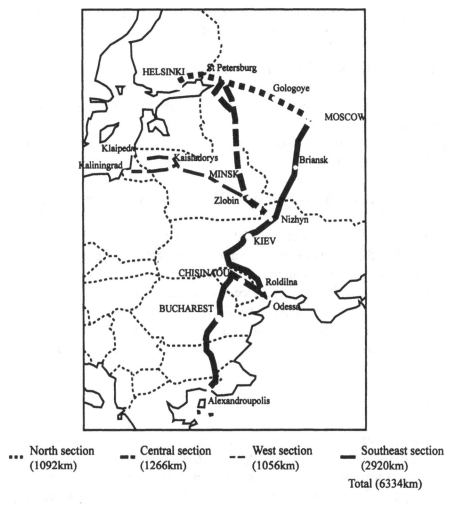

... North section (1092km) -- Central section (1266km) -- West section (1056km) — Southeast section (2920km)

Total (6334km)

Figure 6.1   Rail corridor IX

- *central section (St Petersburg – Nizhyn)* has a total length of 1,266km.

The southeast section (SE) (under study) corresponds approximately to 46 per cent of the total length of corridor no. IX. It crosses six countries (Greece, Bulgaria, Romania, Moldavia, Ukraine, Russia). It also crosses four capitals (Moscow, Kiev, Chisinaou, Bucharest) and serves directly two important sea ports (Odessa, and Alexándroupolis), and one river port (Kiev). Indirectly, of course, it serves a number of other major ports of the Black Sea area (Costanza, Varna, Burgas, and other smaller ones).

In Table 6.1 the most important geometric and operational characteristics of the SE section are given at country level.

Assessing Track Infrastructure and Installations

The assessment of the existing situation as regards the track of the rail infrastructure along the southeastern section, showed that this is characterised by a number of features, which are the following (see also Table 6.1):

The positive aspects are:

- the track is double to a large percentage (2,214km, 75.8 per cent of total section length);
- there is transport possibility of 20 tonnes (t) per axle load along the total length of the line, while for a major length percentage (67 per cent) higher axle loads can be accommodated (22.5t);
- the line is electrified to a large percentage (2,325km, or approximately 80 per cent of the total SE section length). Electrification does not exist in Greece, in Moldavia, and in some small segments in Bulgaria and Romania;
- the traction system is almost uniform everywhere (alternative current 25 KV–50 Hz).

The negative aspects are:

- the superstructure is in a poor condition along almost the total length of the section examined. This fact, in combination with the very 'stringent' geometrical alignment characteristics (basically small curves in horizontal alignment), has resulted in the application of very low commercial speeds:
- there are two different track gauges: normal gauge (1,435 mm) in Greece, Bulgaria, Romania and large gauge (1,520 mm) in Moldavia, Ukraine, and Russia. This fact is a major disadvantage and obviously discourages transportation over the two 'sections' of this corridor.

A substantial part of the length of the corridor does not have electrical signalling (approximately 27 per cent of its total length). Mechanical signalling exists in three out of the six countries crossed by the corridor (Greece, Bulgaria, Moldavia).

Because of the poor state of the track, urgent maintenance and renovation works are executed along many sections of the corridor (especially the sections

Table 6.1   Existing geometric and operational characteristics of SE section of rail corridor IX

Country	Track length (km)	% per country	Track gauge (mm)	Type of track (Single/double)	Load per axle (t)	V max (km/h)	Traction system	Signalling system
Greece	179	6.2%	1,435	179 -	20.0	100	Diesel	Mechanical
Bulgaria	380	13%	1,435	347/33	20.0 22.5	105	Electrical 25 KV–50 Hz (81.5%)	Mechanical
Romania	595	20.4%	1,435	77/518	20.0	160	Electrical 25 KV–50 Hz (77%)	Electrical
Moldavia	209	7.2%	1,520	103/106	22.5	100	Diesel	Mechanical
Ukraine	1038	35.5%	1,520	– 1038	22.5	120	Electrical 25 KV–50 Hz	Electrical
Russia	519	17.7%	1,520	– 519	22.5	120	Electrical 25 KV–50 Hz	Electrical
Total	2,920	100%	1,435 39.5% 1,520 60.5%	706 2,214 (24.2%) (75.8%)	22.5 (67%) 20 (33%)		Electrical (79.6%) Diesel (20.4%)	Electrical (73.7%) Mechanical (26.3%)

Figure 6.2 **The study area: corridor IX, southeast section: Alexandroupolis – Bucharest – Chisinaou – Kiev – Moscow**

from Bulgaria to Moldavia), which impose further low speeds and travel time increases on an almost permanent basis.

In some sections of the corridor, the traffic capacity of the track (because of its poor condition and alignment) has reached its maximum limits and as a result scheduling of additional trains is a problem. This particularly refers to the sections of Ungeny-Chisinaou in Moldavia (capacity covered to 100 per cent) and the Stara Zagora – Mihaylovo and Dabovo – Tulovo sections where current data show that available track capacity is covered 86 per cent and 81 per cent respectively.

The equipment for loading/unloading at land, sea and river terminal stations along the corridor, is not sufficient and in poor standards, having as a result a decrease in the effectiveness and efficiency of the loading/unloading operations.

There is almost a total lack of modern systems for the exchange of information and data (that is, by any modern method of telecommunication such as EDI or Internet-based applications) between the railway networks involved. As a result, operational coordination activities between the networks involved, as well as connection to the customers by this modern means (especially in real time), is virtually nonexistent.

Assessing the Rolling Stock

Regarding the existing situation for the rolling stock, the following points can be observed:

- the rolling stock used in the corridor in all countries is old and needs upgrading to suit modern rail transportation needs;
- all Southeastern European rail networks along the corridor have plans (and the knowledge) to replace or upgrade their rolling stock, but there is lack of the necessary funds or financing packages that would allow such improvements to go ahead. As a result only modest to marginal improvements can be expected in the near- to medium-term future;
- in the Greek and Moldavian sections only diesel locomotives are used, while for the rest of the sections both (electric and diesel) types are used;
- for the Moldavian, Ukrainian, and Russian networks freight rolling stock has been designed taking as primary criterion its 'transport ability' in terms of capacity and weight. As a result, the maximum speeds allowed for this (freight) rolling stock as far as their construction is concerned is in the area of 80km/h (with a maximum of 100km/h).

There are serious operational effectiveness problems in the transfer between the parts of the corridor of different gauge.

Transport Flows and Services

Passenger Transport Services

Regarding passenger service exploitation along rail corridor no. IX, the main observation is that there is a very low level of service with low speeds, low frequencies of service, very poor correspondence of schedules, and low convenience levels for most of the international transports along the corridor.

As regards domestic transportation, there is a different picture, which varies from a positive one for certain sections (Romania, Moldavia) to a negative one (Greece, Bulgaria, Ukraine) as concerns the same criteria mentioned above.

The overall picture for passenger transport flows today is that there are very few truly international passenger flows, and these are low and decreasing in most sections of the corridor. This picture is different when only national sections and trips are considered, but there, too, passenger volumes are in constant decline. The reasons for these rather negative results must be sought in the current state of the economies of the area, which results in low mobility levels anyway, and at the same time lack of the necessary substantial investments for improving rail passenger and freight infrastructure and services to make them more competitive and 'appealing' to travellers.

For international (passenger) transports, the three most influential factors that characterise the current situation are:

1   *Kilometre distances*, for many connections which could justify routing of direct trains (for example, between capitals of neighbouring countries or big urban centres) are generally high and thus they create travel times that (taking into account the relatively low speeds anyway) are non-competitive with those of the other modes of transport (namely road and air). For example the following distances are recorded between capitals or major urban areas along the corridor:

Moscow – Kiev:	878km
Moscow – Chisinaou:	1,578km
Moscow – Bucharest:	2,189km
Moscow – Alexandroupolis:	2,920km
Kiev – Bucharest:	1,311km
Alexandroupoli – Bucharest:	667km

2   *Commercial speeds* are low (and thus travel times are high). These are extremely low. For example the following figures of travel times have been taken from existing timetables, and can be considered as representative of the whole corridor:

Bucharest – Chisinaou:	48.7km/h, 12h32'
Moscow – Kiev:	67.8km/h, 12h50'
Odyssos – Stara Zagora:	33.0km/h, 35h
Svilengrad – Russe:	51.7km/h, 7h15'

The small commercial speeds are mainly due to:
- low speeds, which are imposed due to poor rail track quality, 'narrow' alignment characteristics and delays applied. The average travelling speed in the total corridor length (Alexandroupolis – Moscow) is estimated at below 60km/h;
- delays in nine border stations (on average 45 min, to 1 hour per station) and especially in Moldavia-Romania borders (2 hour minimum due to track change);
- delays due to traction unit change. For the total corridor length there are 7 points of traction change (Mihaylovo, Tulovo, Dabovo, Russe, Iasio, Ungeny, Suhinici).

The routing of a train between Alexandroupolis and Moscow assuming that it:
- crosses the total distance without any correspondence;
- accepts all delays (stop at border crossings etc.), which are defined from present routing and scheduling procedures was estimated to be 56h10' which corresponds to an average commercial speed of 50.8km/h.

In case the stops in certain stations are eliminated, the trip duration can be reduced to 48h (two days), which is equivalent to an average commercial speed of 60km/h (a reduction of approximately 20 per cent);

3 *International train schedules are not coordinated optimally.* In order to demonstrate this problem an analysis was made of the routing of a train between Alexandroupolis and Bucharest. According to current schedules, a passenger train that goes from Alexandroupolis to Bucharest needs 1.5 days for the 667km distance. This total travel time includes a 22h delay (!) for a train change in Svilengrad. This fact alone almost precludes any passengers from using this service.

Freight Transport Services

International and domestic freight data, which were collected from the study team for all the no. IX corridor-crossing countries, are encouraging for all sections of the corridor except for the Alexandroupolis – Ormenio section (Greece).

It would be true to the current situation if we state that *the operation of rail corridor no. IX today (as an overall international corridor) takes place for freight transport only.*

In the normal gauge sections of corridor no. IX, that is, between Rumania, Bulgaria, and Greece, the working team has estimated the total amount of freight moved by rail in 1999 to be of the order of 1,500,000 tonnes annually. This figure reflects the current (economic and other) situation in the countries involved and does not include the national freight movements.

For the future, the potential for increases is certainly there and the picture for (international) freight movement along this rail corridor is one that could triple this figure even as soon as 2005. After 2005 and according to the date that the currently associated countries to the European Union become full members, the picture as regards the demand for freight movement along corridor no. IX can change dramatically from what we see today. This would primarily be the result of 'abolishing borders' along many of the countries in the area of Southeastern Europe, a fact that by itself can have tremendous repercussions and impacts (see also van Geenhuizen and Ratti, 2001). All this is, of course, very promising and points to the fact that the rail corridor no. IX should be upgraded even if the current flows do not seem to guarantee its return on the investment today.

In general, however, traffic flow data for this part of the corridor are scarce and certainly do not include road transport figures. So a reliable modal split figure for total (international) freight movement along the corridor no. IX corridor cannot be given. Our estimate is that road transport has surpassed the 50 per cent figure in the 1990s, and soon will be carrying more than 60 or even 70 per cent of the total international freight transport work.

For the Greek part, the total transport work between Greece and the rest of the corridor no. IX-crossing countries, by all modes, was approximately 7 million tonnes annually in 1999. Of this, corridor no. IX (road and rail) takes only about 30 per cent (2 million tonnes annually) and of this only approximately 10 per cent (220,000 tonnes annually) is moved by rail.

Via the Ormenio station only 50,000 tonnes of freight are recorded per year. This corresponds to approximately only 1 per cent of the total transport work (with all means of transport, from all entrance/exit points), which is produced between Greece and the other rail IX corridor countries. These 50,000 tonnes of freight transport work mainly go to Bulgaria. Transit flows via Bulgaria to Rumania and beyond are negligible.

The Prospects for Corridor No. IX

Introduction and SWOT Analysis

The study made a systematic appraisal of the possibilities and prospects of the rail corridor no. IX, it also formulated a series of technical proposals and recommendations in the form of a medium-term 'business plan' which begins with a SWOT (strengths, weaknesses, opportunities, threats) analysis, that is, the systematic collection and assessment of the strong and weak points of the corridor, as well as its potential for future development and the 'threats' to this development. Then the recommended actions are analysed and presented. These are intended for the high level group that was mentioned in the introduction and which has been set up from high-level rail network representatives of the corridor no. IX railways.

The recommendations were based on the analysis of the current problems and prospect for this corridor but also on a series of interviews between the members of the working team and experts in all countries of the corridor.[3]

They were aimed at producing a first and absolutely necessary level of improvement in order to have the corridor functioning as a credible international rail corridor able to offer an alternative service to the various competing modes and routes that exist today, or that are likely to appear in the future.

The results of the SWOT analysis are given in Table 6.2. As this table shows, there are a number of opportunities for the development of this corridor. The biggest one lies in the fact that all countries along the corridor (with the exception of Greece) have a long and very strong tradition in favour of using railways. In the 1980s the railways carried more than 80 per cent of all freight (and passenger) flows, and in spite the current decline in view of the strong competition from road, the prospects for the revival of rail are there and are good.

To this end a strong ally will be the prospects for European Union membership for Rumania and Bulgaria, and the importance attached to this corridor as a TEN corridor of pan-European importance. It is characteristic that in its new White Paper for Transport Policy (European Commission, 2001) the Commission pledges its full support to the development of these networks in the east. A third 'opportunity' as mentioned in the SWOT analysis lies in the expected congested sea transportation via the Bosporus and the Dardanelles today, which is expected to create further interest for this corridor as an alternative route.

Table 6.2   SWOT analysis for freight and passenger transport via axis no. IX

Strengths	Weaknesses	Opportunities	Threats
Rail as means of transport has been in the past and continues to occupy at present an important position in the economies of Southeastern European countries	Operational problems (different track gauges, many border points, different gauge)	Infrastructure and development works are taking place in the port of Alexandroupolis, which is the end of the axis, that will allow it to have a proper transfer point for loads via this rail axis and its associated rail network	Road transport continues to gain an ever-increasing share in freight and passenger transport
	Low commercial speeds		There are other competitive rail axes that have been given higher priority at national programmes
The geographical position of axis no. IX gives the possibility, under certain cost conditions, for credible alternative land transportation links instead of the sea transportation via the congested Dardanelles passage	Low levels of investments on this axis	OSJD has achieved very good cooperation levels among its members (SMGS agreement)	Construction of the road axes east/west gets higher priority in relation to rail (e.g. the construction of the Egnatia road axis, or axis no. VIII)
	Bad quality of rail superstructure	The expected rail reformation in Europe gives new opportunities for change from road to rail transportation	
	At present the port of Alexandroupolis is not connected to the axis		Proposed oil pipeline connecting Varna to Alexandroupolis may diminish any potential for oil transports via this axis, while at the same time it will make the sea transport via Bosporus less congested and thus more attractive
The rail IX axis has several ports 'connected' to it via short rail links or directly. All countries that are now 'promoting' rail axis no. IX have good relations and cooperate in many fields	Some of the sections of the IX axis are not primary links in the rail networks of their respective countries and have a low importance and priority in the respective funding	Predicted economy development of Southeastern Europe countries will allow increase of transportation work	
		Congested sea transportation via the Bosporus and the Dardanelles today	

Recommendations for the 'Normal Gauge' Sections

All countries of the southern part of the corridor have put in place or are starting to execute plans and programmes for the improvement and modernisation of their respective rail networks. However, these are almost exclusively focused on their national priorities and programmes and there is little evidence that the international characteristics of the corridor are taken into account. In other words, one misses the sort of interventions that are necessary in order that the corridor operates efficiently and not with a minimum level of service common among all implicated networks.

So the first and probably most important recommendation, which was taken up by the current working group, was that the rail networks in all countries of the corridor should proceed to a systematic listing of priorities and interventions that are currently planned in each network, and *discuss and agree on coordination of actions and investments*.

This coordination should be the first priority in the Agenda of the Corridor IX High Level Rail Committee mentioned in the introduction of this chapter.

A number of other actions and projects for the normal gauge section (Greece to Moldavia) were also recommended. These actions have both long- and short-term perspective. However, those for the longer term must be established as 'common objectives' and parts of a long-term master plan for the revitalisation of the corridor no. IX. Once agreed by all networks concerned, they should systematically be applied in their national rail network planning and development works. The most important recommended actions in this sense are:

1 all rail infrastructures along the axis should be upgraded to have as a *minimum* weight the 20t/axle load and *minimum* speeds of 120km/h. This configuration is recommended as an acceptable balance between efficiency and economy, that is, one that would speed up freight trains and ensure a satisfactory level of service for rail freight transport while at the same time keeping costs at a reasonable level;

2 for passenger transport services (both at national level or between neighbouring countries) higher speeds of the order of 140–160km/h should be aimed for. This aim is likely to require higher levels of investment for all countries except Rumania where the required changes to reach this objective are not so substantial;

3 a *strict* maintenance programme, common to all networks, must also be devised and agreed upon so as to maintain the above speeds;

4   installation of electrification along the total length of the corridor in the considered section (of normal gauge) is also necessary. This will involve approximately 30 per cent of the current length (or approximately 900kms). The electrification programme must again be coordinated so as to achieve maximum compatibility between the countries involved. Its realisation would ensure operational ability, synthesis of longer trains and an increase of commercial speeds;

5   installation of electrical signalling along the total route length. This intervention regards basically Bulgaria where most of the needs are at the moment. It will contribute to make the route more secure, increase track capacity, and commercial speeds;

6   increase track capacity in certain 'centrally' located parts of the corridor, which by their limited capacity 'affect' many other sections, the parts being *Stara Zagora – Mihaylovo,* and *Dabovo – Tulovo* in Bulgaria;

7   promotion of fully intermodal transport services, that is, cooperation with other modes to provide door-to-door services for freight, and (optimally) freight logistics integrated over the whole transportation chain. These are the current state-of-the-art services that the new fully competitive market requires (see, for example, Brewer et al., 2001). They should involve, at least, rail and road as well as rail and sea transports and will require development of appropriate IT as well as physical infrastructure, for example for modal transfer facilities at ports and railway terminals. This should be seen as an objective to be pursued both in the short and the long term in order to 'secure' the *position of the corridor vis-à-vis road transport and as an alternative to the all sea transport via the Dardanelles and Bosporus;*

8   improve by way of priority the rail sections that give access to the sea and river ports served by the corridor and the modernisation of their equipment (concerns mainly the ports of Varna, Burgas, Kostantza, and Alexandroupolis);

9   the rail station in Dimitrovgrad must be upgraded soon as it is an important nodal point of axes no. IX and IV. The station is under reconstruction and upgrading but the provision for combined transport installations is lacking. More generally, the section Svilengrad – Dimitrovgrad in Bulgaria is of particular importance as it belongs to both the corridor no. IX and corridor no. IV;

10   modernisation of border station installations. These will aim at developing a minimum of modern infrastructure available at all border crossings. The objective should be to provide *a minimum* of modern building facilities, for example, for passenger checking services (these can be used as offices

for the customs, police and other services when there will be no need for stopping trains to check passports, etc.) as well as IT (information technology) infrastructure. In another cooperation scenario case, *common building facilities* for both sides of the crossings could be planned;

11 installation of common (or fully compatible) electronic systems for the exchange of information and data between the rail networks of the corridor. This should be established by way of priority. Such systems for the exchange of information, based on EDI or web technology, are becoming widely used in western European railways today, and they prove their value in terms of time savings due to better schedule coordination, faster controls, better information to the customers, etc;

12 formation of common tariffs and tariff policies for international transports via corridor no. IX. These 'common' fares will have the meaning of competitive fares against road transports, which will not be the sum of the corresponding national sections, but they will be defined on a door-to-door basis and competitive with the corresponding tariffs to be found by the customers on the similar route by other modes or services.

13 a common and concerted attempt must be made for the drastic decrease of border delay times and simplification of procedures. A first reduction of wait times at borders of below 50 per cent of the present wait times at border crossings should be effected within a time frame of the next two to three years. This can be done primarily by bilateral agreements between each network (and between corresponding countries) and could 'target';

- *passenger transports*, by eliminating or drastically reducing passport, police, and customs controls at borders by performing them in motion in the train at the preceding section, by use of modern PCs and telecommunications technologies;
- *freight transports*, by checking only the documents for each wagon which could be sent to the border station in advance via EDI, and performing all the necessary controls and inspections at the starting and terminal points of the trip. Again bilateral or multilateral agreements will be necessary.

Recommendations for the 'Large Gauge' Sections

These apply to the sections which cross Moldavia, Ukraine, and Russia.

Regarding the infrastructure the critical question that concerns this section is the upgrading of the installations for gauge change from 1,435 mm to 1,520 mm at the Romania-Moldavia borders. This should be a major subject for consideration by the working group of the corridor no. IX railways. Experience in the Baltic countries on this subject suggests that cost effective and technically sound solutions are possible.

Other suggestions include:

1   upgrading of rail infrastructure and securing, at minimum, axle loads of 22.5t and speeds of 120km/h should be aimed at in this section;
2   application of a rigorous maintenance programme, which will secure the above speeds;
3   installation of electrification along the total length of routes in the Moldavian section. This intervention will improve operational ability, synthesis of longer trains and increase of commercial speeds in this critical section of the corridor;
4   provision of combined transport trains along the total length of corridor but primarily in connection to the port of Odessa;
5   installation of electrical signalling along the total length of the route in Moldavia. This intervention will improve track capacity and traffic security in this critical section of the corridor and will increase commercial speeds;
6   increase of track capacity in the sections of the corridor with high traffic loads and especially at the Ungeny – Chisinaou – Bendery section in Moldavia;
7   improvement of rail access at sea and river ports and modernisation of their equipment (Odessa, Ilyichevsk, Kiev).

As regards the *exploitation*, the same recommendations made for the 'normal gauge' part of the corridor apply here as well.

Conclusions

The need for appropriate improvements in rail infrastructure and operation in order to provide a credible alternative to road transport, as well as coordination with the other modes of transport, is ever pressing and present in Southeastern Europe. The corresponding section of corridor no. IX in this area should become a priority section to improve since as it is today it hardly merits the definition of an 'corridor'. For progress to be made, a number of well-coordinated actions are necessary at many levels. These have been presented in a summary form in the immediately preceding sections. They are already important in that they show the spectrum of actions necessary, and the details given at the study can be used as a first approximation of a master plan.

The countries and railway networks of the area must be helped in terms of coordinating their efforts, and financing the development of this infrastructure. This can best be done by formulating, agreeing, and implementing as soon as possible a 'master plan for the development of the rail corridor no. IX'. This master plan should have a timeframe of 15 years, that is, 2015 as its target date; besides its priorities and functional specifications of the necessary works, it should also define the necessary financing instruments.

Then there are many administrative actions that must be instituted with the proper cooperation by all governments in order to facilitate national and international transport along this corridor. Perhaps the most urgent of these actions are the ones towards easing the difficulties in crossing borders and introducing modern information technology in communications between the rail operators and between them and their customers. These are two areas where a lot of improvements can be made in a relatively short period of time, and it is strongly recommended that they are given full priority.

By increasing their level of cooperation and common planning of their investments, the railway administrations along corridor no. IX can ensure that maximum weight is imposed on their demands for financial assistance and at the same time they can coordinate their actions to introduce the low cost short term improvements suggested in the previous paragraph. It is hoped that the existing high level coordination committee of representatives from all the railways involved will try to initiate this level of coordination as a first stage, to a more institutionalised cooperation at higher levels.

Notes

1 The Laboratory of Transportation Engineering of the Aristotle University of Thessaloniki, on behalf of Greek Railways, conducted the study.
2 This was done by utilising the services of the Southeastern European Transport Research Forum (SETREF) and its network of researchers across Southeastern Europe.
3 The contribution of the SETREF network (Southeastern European Transport Research Forum) is gratefully acknowledged in this respect.

References

Aristotle University of Thessaloniki (AUTh) (2001), 'Pre-feasibility Study for the Development of the Railway Axis No. IX', *Greek Railways*, January.

Brewer, A.M., Button, K.J. and Hensher, D.A. (2001), *Handbook of Logistics and Supply-chain Management*, Pergamon, Oxford.

European Commission (2001), 'European Transport Policy for 2010: Time to Decide', White Paper, COM(2001)370, September .

International Union of Railways (UIC) (1994), *Infrastructure Projects in Central and Southeastern Europe*, December.

Van Geenhuizen, M. and Ratti, R. (2001), *Gaining Advantage from Open Borders*, Ashgate, Aldershot.

Chapter 7

Major Improvements and Challenges in Transport Logistics along Corridor V in Hungary

Péter Rónai and Katalin Tánczos

After the temporary economic decline in middle Europe during the 1990s, a continuous sustainable recovery process started in most of those countries which had previously had a planned economy system. Industrial systems and their products and markets changed suddenly, raising new technological demand for cost-effective infrastructure investments. This chapter aims to give an introduction to the most important developments, plans and future needs which appeared in Hungary along the fifth European transport corridor during recent years. In order to understand the specifications of the chapter later, we start with a general overview of the economic and transportation-economic trends and difficulties of the whole state. Recent data on transport demand and modal split will be then examined. Improvements in the field can be classified into two different groups: link development and conjunction point development. The next section provides a 'virtual trip' along the Hungarian section of the fifth European transport corridor. All significant junctions and nodes will be examined: the Záhony trans-shipment area (the biggest 'offshore' port of Hungary), motorway constructions M3 and M7 (the largest road construction works of the last decade); the relevant logistic centres (BILC, Székesfehérvár) and the new railway track line connecting Slovenia with Hungary. Planned and accomplished infrastructure development aim to promote the cooperation of more transportation modes in the freight sector. The next part of the chapter gives exact data and shows the cost efficiency analysis tool developed at the Department of Transport Economics (BUTE). The methodology shown describes how the department contributed to large-scale infrastructure projects, and provides the financial feasibility data of the planned projects. Input and output values are defined here and a compressed output diagram is presented to make the model more comprehensible. The long-term financial study shows the feasibility of the 'public-private partnership' system of the whole multimodal terminal. The importance of the social-economical efficiency assessment and the possibility of internalising external costs of the construction are underlined.

Introduction: General Macroeconomical Overview

The political and economical situation of Hungary showed much similarity with other Middle and Eastern European countries. After the change of political system in 1989, the much more difficult work of changing the structure of the economical environment started. The tasks involved were similar to those in neighbouring countries: an old and ineffective planned economy system had to be turned into a competitive market economy by changing the output constitution and the marketplaces themselves. All this proved to be a hard task for the country. At first, the GDP decreased suddenly, the lowest value occurring in 1994, at about 80 per cent of that produced in 1989. After the middle of the 1990s a slow economic recovery started, reaching a comparatively fast growth rate at the time of writing of more than 5 per cent a year. This made it possible to overcome the temporary difficulties of the system change and to put production on an expansion course. The total output of 1989 was reached again in 2000, but with an already thoroughly different economic structure: 80 per cent of total export of the country was transported to the European Union states. Industrial production is mainly driven by foreign working capital, especially in the field of heavy machinery and production equipment.

All these structural renewals led to significant changes in the goods transport sector. Share of bulk products decreased suddenly, but the need for reliable logistics services emerged. As a result of this, the market share of railways has fallen from 70 per cent to 30 per cent, and the overload on roads started to become increasingly unbearable. The development of transport systems and infrastructure was not parallel with the overall economy performance. Already by the closing years of the 1980s Hungary had a rather low quality road and railway network. As governmental sources of financing were reduced, even the maintenance of state-owned transport lines and tracks was neglected. This led to a noticeable fall in quality in all transportation fields. Today, one of the greatest tasks of road constructors and maintainers is to reach the 11.5 tonne axle load limit (in Hungary now only 10.5 tonne is allowed) to comply with the European Union standards. Construction of motorways and high-speed roads also has to be accelerated. When forming an opinion about railway transportation in Hungary, similar statements can be made. Hauling power in particular is well below the desired technical level. The Hungarian State Railways (HSR) are not in possession of multi-current-type engines, therefore the requirements of interoperable border crossing can not be fulfilled properly. Although the state railway company has many freight wagons, they do not meet the new expectations, and are not capable of carrying trucks in Huckepack RoLa (rolling road) trains. A further matter is that tracks do not always allow the operation of high-speed trains, therefore passenger transport quality indicators cannot reach the planned level. Unfortunately, the development of road and railway infrastructure and railway machinery equipment lag behind the general economic growth. A significant upswing could only be noticed during the last couple of years. One direct consequence of the above-mentioned changes in the early 1990s was the governmental initiative. By 1992 places of planned regional and international logistic centres had already been defined, some of which are partly in operation today. As the preferred European corridor routes were specified in Crete and in Helsinki (1997), further concentrated state and private developments took place along the desired lines.

In this aspect Hungary has a special position. Despite the fact that the area of the country is rather small (93,000 km²), four European corridors cross within the country (IV, V, VII, IX/b). The fifth one – the previously mentioned Helsinki corridor – is of great importance, as it connects southern parts of Western Europe and middle part of Eastern Europe. It provides further connection to the east through Ukraine.

Figure 7.1 shows the territorial set-up of Hungary and the main routes of the fifth Helsinki corridor through the country. There are other branches of the corridor passing the country, but they had only marginal importance. The thick black line on Figure 7.1 gives the path of the railway line of the fifth corridor, the broken line provides the Hungarian section of the road lane of the corridor. There are several places numbered along the line: these will be discussed in this chapter.

The western end of the fifth corridor is Venice and Trieste (Italy), the eastern endpoint lies at Lemberg (Ukraine). The whole length of the corridor is approximately 2,800 km, of which the Hungarian railway section is 1,040 km. The road section is 630km long. The PHARE study of the corridor includes the projects of 44 planned road developments, 48 railway developments and three combined transport developments, of which 16, 13 and one respectively are located in Hungary.

One aim of this chapter is to introduce the most important developments and plans appeared along the fifth European transport corridor during the last decade. The already started or finished developments along the corridor line

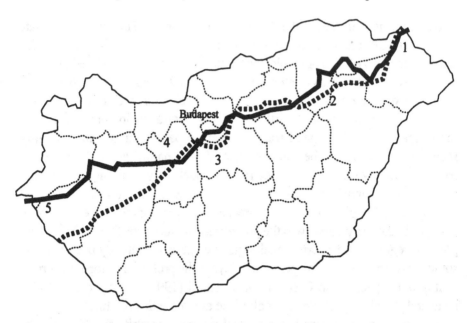

Figure 7.1   Map of Hungary with the paths of the fifth Helsinki corridor

will be discussed, and reference to each numbered project along the corridor will be made. From the viewpoint of the accession countries, attention is drawn to the infrastructure developments in close connection with the logistic services in order to make potential investors aware of profitable regions of Hungary. A further aim is to underline the feasibility study specifications in the country, handled by the presented model.

Transport Volumes and Modal Split Data

Parallel with the general changes in Hungary's economy, a restructuralisation process commenced in the transportation sector. Market share and volume of bulk products decreased suddenly at the beginning of the 1990s. According to the economic reorientation, the direction of shipments was balanced from east to west. The value share ratio of the economy exchange (more than 70 per cent of all foreign trade is between Hungary and the European Union countries) is not exactly represented in the transported amount of tonnes, because a rather high volume of energy sources (for example, fossil oil) and heavy industry raw materials (such as iron ore) is imported from the CIS countries.

Table 7.1 shows Hungary's export and import transport volumes from 1993 to 1999.

Table 7.1 Foreign trade volumes of Hungary without waterborne transport

	Import (1,000 t)	Export (1,000 t)
1995	19,234 1	6,863
1996	14,326	17,620
1997	17,681	17,595
1998	19,032 *	18,556 *
1999	17,868 *	19,174 *

* Contains linear extrapolated data.

Source: Institute for Transport Sciences Ltd.

That export volumes are nearly the same as import volumes is a result of the good market position of Hungarian agricultural products in the European markets. On the import side, the balance is provided by energy sources and heavy industry base materials. The value balance of foreign trade is not as good: the value of imported goods is usually higher than that of exported goods, and this results in an ongoing trade deficit.

Regarding the forecasts for the next decade, a slow growth is predicted (shown in Table 7.2).

Table 7.2 Transportation volume forecasts until 2010

Year	Export (1,000,000 t)		Import (1,000,000 t)		Transit (1,000,000 t)	
	2005	*2010*	*2005*	*2010*	*2005*	*2010*
Road	9.1–9.4	9.7–10.1	5.2–5.3	5.5–5.6	7.8–8.9	18.0–9.5
Rail	11.6–14	11.8–14.9	16.3–16.4	16.6–16.8	2.2–4.6	2.0–5.4
Waterborne	3.1–3.0	3.1–3.2	1.3	1.3–1.4	5.1–5.2	5.9–6.0
Total	23.8–26.7	24.7–28.1	22.8–23.0	23.6–23.7	15.1–18.8	16.0–20.9

Source: Institute for Transport Sciences Ltd.

Uncertainties arise from the doubtful growth rate of the economy in general, the progress of transport infrastructure development, and the global economic situation. Three important logistic junctions were mentioned in the abstract along the fifth European transport corridor: Záhony, Budapest – BILC and Székesfehérvár. These will be discussed in more detail later. Table 7.3 gives an introduction to the road transport volumes in the effect field of the three logistic providers.

Table 7.3   Road transport volumes in the circumstances of the three logistic centres along the fifth corridor (1998)

Service provider	Export (1000 t)	Import (1000 t)
Záhony	267.1	161.1
BILC	2,467.5	2,341.9
Székesfehérvár	797.9	331.7
Total	3,532.5	2,834.7

Source: Institute for Transport Sciences Ltd.

When considering the modal split of bulk and heavy machinery products between road and rail, the ratio is not as bad as the previously discussed overall value. Still, a more environmental friendly and more economic ratio than the current one (see Table 7.4) should be reached. In both the export and import columns the sum of the two neighbouring numbers give 100 per cent. As the table shows, modal split data show a quite advantageous ratio when considering goods with a high weight. The ratio when treating all goods of the economy does not look as good.

Záhony-Port Trans-shipment Area

Although Hungary does not have a coastline, it does have a port: the Záhony ashore port, working as a turntable for railway deliveries between east and west. Trans-shipment because of the railway track gauge has been carried out in this region since 1948.

The land trans-shipment complex, with an area of 84 km², referred to as the eastern gate of Hungary, includes Záhony and the eight railway stations in its direct neighbourhood. In Figure 7.1 the trans-shipment area is indicated by the

Table 7.4   Modal split of transport of bulk products and heavy machinery

	Export		Import	
	Rail %	Road %	Rail %	Road %
Agricultural products	69.85	30.15	75.61	24.39
Foodstuff	36.37	63.63	59.24	40.76
Solid fossil energy sources	100.00	0.00	99.17	0.83
Fossil oil and its derivatives	86.21	13.79	79.08	20.92
Ores	4.03	95.97	100.00	0.00
Metal basic materials	60.11	39.89	77.59	22.41
Fossil- and building materials	65.50	34.50	73.50	26.50
Fertilisers	25.01	74.99	100.00	0.00
Chemical products	36.99	63.01	74.05	25.95
Heavy machinery products	40.19	59.81	62.39	37.61
Total	49.54	50.46	78.21	21.79

Source: Hungarian State Railways.

number 1. Its main activity is the trans-shipment of consignments arriving from the Community of Independent States (CIS) and adjacent countries through the border stations of Csop, Ukraine (Záhony) and Batyevo (Eperjeske), in wagons with wide track gauge (1,520 mm) into railway wagons with normal track gauge (1,435 mm) and their transporting to and through Hungary to the other countries of the continent.

Here, in addition to the tasks of a railway border station and trans-shipment the storing and handling of several million tons of goods, in accordance with the demands of clients, are carried out annually. Regular investments have made it possible for the port of Záhony to have the capacity for the trans-shipment of 16 million tonnes of goods and for the further delivery approximately 1.5 million railway wagons. Unfortunately, the large capacity is not currently utilised: not more than 5.5 million tonnes of goods are transported through the port annually. The centre of the trans-shipment region is Záhony and the individual areas of trans-shipment at various stations were created and adjusted according to the type and the physical characteristics of the goods transported.

At the bulk goods loading areas, trans-shipment of shipments of grains and powder from wide self-discharging wagons into open or normal railway wagons is carried out, These can be filled from the top either by gravity or mechanically. Iron ore and coal are not susceptible to changes in the weather, therefore the unloading of the wide wagons is performed by open-air gravity conveyors directly into the railway wagons or the spoil area. Discharging is promoted by vibrating equipment. Mechanical filling is done from the open wagons by large volume scoop shovels mounted on cranes or mobile loading machines. At the covered platform the loading of artificial fertiliser and grains drained into the bunker is undertaken by swift conveyors into the normal wagons.

Since it is likely that Hungary will become a member of the European Union in the near future, eastern state border crossings and technical equipment are important for the union itself as well. This is the reason for Western capital investments, which transform the old trans-shipment area to an international logistic centre with continental strategic tasks. Modernisation of the technical facilities has already started, and Záhony is a preferred area for governmental as well as European Union investments. If all the processes which have been initiated continue in the desired way, Záhony will be one of the most significant chain loops between the European Union and the CIS countries.

Motorway Developments towards the East

According to government calculations, the construction of 50km of motorway is followed by the GDP increase of 0.5 per cent annually. This number does not only reflect the well known multiplication effect of the transport infrastructure, but also the improvement in economic competitiveness of the region. This results in more foreign and domestic capital investments. Another interesting number is the reduction in the unemployment rate: if the access time measured from the capital (or the western state border) of a certain region is decreased by 10 minutes, the unemployment rate falls by 1 per cent. Apart from the corridor priorities, these have been the most important reasons why the government decided to accelerate the existing motorway development plans.

A first result of this is the new section of the M3 motorway from Füzesabony to Polgár (approx. 67 km). The section is indicated with a number 2 on Figure 7.1. Total investment cost is HUF 69.5 thousand million (approx. € 278 million). A new bridge over the River Tisza is also part of the construction. Northeastern Hungary was the most adversely-affected region after the political system change: former Soviet-type heavy industry no longer had a market, unemployment ratio increased suddenly and the area was not attractive for foreign capital because of the poor accessibility. It is hoped that the economical, social (and sometimes) moral decline of the region is now over, as the new motorway investment changes the industrial set-up as well.

The construction of the above-mentioned section was due to be finished in November 2002.

The regional logistics centre is expected to be established after the completion of the new motorway. The location of the centre is still not decided. Former heavy-industry junction Miskolc and the smaller, but rather fast-growing, industry town Tiszaújváros are in competition for the centre.

As Hungary's supply of motorways and high-speed motor roads is well below the desired level, further developments are needed to reach the average ratio of the European Union countries. Six hundred kilometres of motorways and high-speed motor roads are expected to be finished within the coming 10 years.

Hungary's Greatest Logistic Centre

Some 150km to the southwest, the Hungarian capital Budapest 'pulsates' as the heart of the country: the city is unusually large compared to the area of the country. The capital and its agglomeration provides 60–65 per cent of the nation's GDP. Therefore special needs in the freight movement field have to be fulfilled, too.

One of these needs is the large amount of transit freight traffic passing Budapest. As mentioned in the introduction, the capital lies in the intersection of four important Helsinki corridors. Regarding transit traffic, a star-like transport net structure concentrates the transit traffic of the whole Hungary into the overloaded capital. The other need in the freight transport field is the local distribution for the 2 million people who live in the metropolis. Thousands of small shops and many dozens of hypermarkets lay claim to goods transport.

The Budapest Intermodal Logistic Centre (BILC) will be the largest logistics centre in Hungary, its full capacity is planned for the transfer of more than 150,000 TEU (20 ft equivalent unit) annually. The rough building place is indicated by the number 3 on Figure 7.1. It has been supported by the European Union because it will play a significant role in central Europe. The centre was under construction in 2001, and will work as an international multimodal distribution terminal in central Europe. Resulting from the above-mentioned tasks, it will have double duty: it will serve as a turntable between road and rail for incoming, outgoing and transit traffic and, at the same time, it has to fulfil the requirements of the city logistic services.

The centre will take over the tasks of the overloaded RoLa terminal in Budapest, and provide a close connection to the Free Port Csepel, allowing shippers to use the benefits of all three transport modes. Furthermore, it makes it possible to establish a close connection to the airport at Ferihegy (both for urgent shipments and operation crew transport).

The BILC project is a promising cooperation of private and state investments. Public-private partnership is needed to allocate the required amount of money to the construction. Total investment amounts to HUF 10 billion (approx. €40 million).

Motorway Developments towards the West

Foreign capital investments arrived unbalanced to the different regions of Hungary during the last ten years. The Transdanubian (west from the river Danube) was much more liked by western firms and investors than the eastern half of the country. As a result of this, a sparkling industrial life started to develop in the region, despite the lack of adequate educated manpower or of good quality transport infrastructure.

Thanks to the unexpectedly fast changes of the economy in western Hungary, the 'capital' of the Transdanubian region, Székesfehérvár, became an industrial, cultural and administration centre of the area. The town has one of the most significant logistic terminals of Hungary. The logistic service provided at the terminal gives a good response to the needs of many multinational firms and enterprises settled in the region after the beginning of the reorientation.

In the late 1960s–early 1970s, the M7 motorway was built as far as Székesfehérvár, and later to Balatonaliga. A two-lane motor road extends the route to Zamárdi. Number 4 indicates the section of the M7 motorway in Figure 7.1. The old motorway construction technology, based on concrete application, together with imperfect execution, resulted in a pretty low quality road surface by current standards. The work of fixing the seesaw concrete blocks and covering the surface with high duty asphalt started at the end of March 2000. The reconstruction will be finished by the end of 2002 with the widening of the motor road to a motorway from Balatonaliga to Zamárdi. The total investment cost is approximately HUF 35 billion (€140 million) for the 110 km-long section. An extension of the line to the state border was due to start before the end of 2001, since the Croatian and Slovenian sections of the fifth corridor are mainly covered by motorways, which will soon reach the Hungarian state border.

Building the Hungarian-Slovenian Railway Connection

The first railway connection of Hungary with Slovenia was established in 1906. After the Second World War transborder traffic was prohibited, and the local railway traffic was given up in 1980 when the decision was made to liquidate the track itself from Zalalöv to the state border (the line is indicated by the number 5 on Figure 7.1.).

After the given corridor priorities in Crete and the political changes in the former Yugoslavia, the restoration of the railway connection with Slovenia has been decided upon. Slovenia was the only one neighbour country in the 1990s which did not have a railway connection to Hungary. In the mid-1990s only money necesary for expropriation of the land was available. Since the plans were in according with the strategic aims of the European Union, obtaining European capital for the construction was possible. The German government gave DEM 120 million (HUF 15.6 billion) credit, while the PHARE programme contributed €10 million (HUF 2.5 billion). Construction work started in 1997.

Soil quality is special with low possible surface load, and rare species are living in the area, this is why special construction technology was used. The 18,5km long track is not only a pure railway connection, it means one of the most modern results of technology developments. At Nagyrákos the longest viaduct of Hungary was established with its 1400m length. Another viaduct has been built with 200m length. Five new stops have been constructed and the existing station at Zalalöv was renewed and extended. A new tunnel with 350m length was built near Nagyrákos. Railway operation started at the end of 2000.

The Support Instrument for Financial Planning and Implementation of Large Scale Infrastructure Projects

The presented infrastructure investments are only a part of those established and planned in Hungary. The hastened growth of the Hungarian economy and the need for accelerated accession to the European Union necessitates appropriate tools for evaluation infrastructure projects.

Among these is the provison of feasibility studies of the planned transport development investments and reconstruction, together with the application of the complex efficiency assessment procedures and methodologies already generally used for consideration and preparation of financial solutions on a national economic level in European Union countries.

Getting deeply acquainted with solutions used in the everyday practice of European Union member states and the systematic application of them is justified by the need for definition of the role and function of the state in the central Eastern European countries after the gradual development of market economy to comply with the new conditions. This made it necessary to transform the approach and data contents of previously applied calculation procedures.

At the same time in the region the central budget sources have chronic deficit, and arising from the demand for restructuring national economic systems in wide approach, competition for external capital sources is increasing.

All these changes require efficiency assessment on national economy level fitting the new aims and databases related to this, in the field of large size transport infrastructure developments of significant effects, serving also public interests.

Usually two assessment methods are adopted for evaluation research and comparing different development versions in the field of transport infrastructure:

- cost-benefit analysis (CBA); and
- multi-criteria analysis (MCA).

Cost-Benefit Analysis

The cost-benefit analysis takes the income, investment and operation cash flows into account during the whole lifetime of the project in such a manner that in these cash flows beside the merely monetarised values it registers also the monetarised values of many other effects concerning public interests. This assessment technique needs the provement of concerned effects in monetary units (e.g. to consider the travelling time effect, the value of time saving due to realisation of the project is needed to be expressed in monetary values).

The cost-benefit analysis is used for deciding whether the social benefits expressed in monetary values exceeds the monetarised social costs, or not. This solution makes possible for decision-makers to answer the question concerning the social net income of the examined project during the estimated project life cycle (or the extents of sacrifices needed, because of general establishment loss).

Cost-benefit analysis used for efficiency evaluation exclusively for public (state) financed transport infrastructure projects makes it possible to find out the contribution of the certain capital investment to other, in the wide sense, realised aims, beside the financial profit. This procedure is suitable (appropriate methods taking into account) to analyse joint financed (public and private) cases, too.

Applicability of the cost-benefit analysis depends on the possibility for expressing the effect of the project realisation in monetary values. Hard to monetarise are for example environmental damage effects, accidents, and psychological effects.

Multi-criteria analysis

Multi-criteria analysis has been developed for analysing cases, which do not make it possible to take all effects into account with their monetary values. Using this method, decision-makers' opinions do not have to be based on monetary values, they can contain objective figures. Multi-criteria analysis is usually used in cases where more project variants supplied with not only monetary values have to be ranked in order to select the best one of them. As a result of this, by the MCA methods is necessary to have more alternatives available (but at least the two cases of one possible establishment and the status quo case), and for each, non-monetarised alternative the summarised value of utility functions represents the considerations of the known assessment.

Using this method, in the course of the solution, with monetary values expressed criteria are usually transformed to dimensionless utility functions, in order that nothing can stop to pull them together with the non-monetarised ones. The priority order issued from this procedure is unstable because of the many subjective units included in the algorithm.

Operation of the Developed Complex Assessment Model

Causes mentioned above, and further European Union requirements, promoted the development of a complex model-system analysis at the Department of Transport Economics, Budapest University of Technology and Economics. The system beside is capable of taking the special features of the Hungarian transport infrastructure investments (relatively high inflation rate, weak local purchasing power, lack of reliable traffic data) into consideration. Furthermore, the system is able to:

- handle all mentioned objectives flexibly;
- calculate other important indexes, which various actors (investor, financier, state, national and international banks) can be interested in;

- calculate from the point of view of the feasibility of the project significant financial, economical and (taking the external effect into consideration) 'quasi social' refundation rate.

From the overview and analysis of the most important features of projects appearing in the practice and the methodical requirements of the above described efficiency assessment conceptions it seemed to be practical to develop a model which meets the needs of the most complicated tasks. Besides all this the programme system makes possible to carry out sensitivity analysis which helps to simulate 'external' (macro-economics indexes like inflation, interest rates, production- and consumer price index, foreign exchange rates), income forming (e.g. traffic volumes, tariffs, fees, etc.) and 'internal' (execution time schedule changes, applying subcontractors, energy, fuel, wage and its charges price change, applying newer technologies, realisation with changed technical contents, etc.) factor changes in connection with the model and it is possible to analyse and test the effect of these changes in time concerned to the realisation of the project.

Input and output data of the model can be classified as follows

- input data and data groups:
    - macroeconomics data (inland and proper foreign price index forecasts, similarly interpreted and specified – e.g. construction industrial – producer price index forecasts, typical interest rates, exchange rates of proper relations, economic control in the field of taxes, depreciation, capital raise/decrease, apport calculation; prevailing margins and interest rate basis of deposit- and credit accounts);
    - source structure features (specification of own and foreign /credit or support/ sources, seniority grade, sum volume, foreign currency name, preferred order of calling in, charges connected to the given source, profile of capital instalment);
    - income forecasts (specification of possible income sources and denomination of them in categories, mechanisms for index derivation of incomes);
    - data of capital expenditures (name of each capital expenditure, possible categories, build volumes and flat rates, mechanisms for index derivation of capital expenditures);
    - data of operation expenditures (appearing categories, volumes, specific constant and changing costs, mechanisms for index derivation of operation expenditures);

- output (computed) data:
  - data of the cash-low table (the task is to ensure positive or at least zero balance in all time intervals, therefore the ground is to calculate interest instalments, taxes and achievements before depreciation from the current operation expenditures and incomes, moreover, to specify from the input control data the liable taxes. Financial cash-flow can be determined by the balance of, in the examined time interval, appeared expenditures in connection with finance structures and the applied sources);
  - financial indices (internal rate of return, capital refund index, credit and project run debt recovery ratio and interest recovery ratio for the whole credit amount and for primary credits, inferior credit staff change, incomes and expenditures of the state when about a state realisation, state guarantee duties and doing these duties).

In order to obtain the mentioned environmental assessment, the external environmental cost is desegregated by vehicle types and technologies, geographical scale of impacts, and the type of infrastructure.

The scale of the impact of environmental cost categories is very different, both in space and time. Whereas airborne pollutants are mainly a problem at the local and regional (e.g. European) scales, the effects of greenhouse gas emissions are global in nature. Noise impacts are restricted to the very local scale of several hundred metres from the emitting source. Impacts on nature and landscape are restricted to the range of several kilometres from the cause, as well as soil pollution. Water pollution on the other hand may affect areas in the range of up to several hundred kilometres. The same is true for nuclear risks, which in case of an accidental release may affect all of Europe.

When considering the socioeconomic consequences of an infrastructure project, the model leans on to the following aspects:

- *efficiency*: the first reason is to ensure an efficient way of society's resources allocation across modes and between transportation in general and all other goods and services. Since not all costs are currently fully covered by the user of a certain mode (e.g. environmental), there is a tendency to make overuse of one mode and under-use of others. The user should pay the cost of the service as long as its value is at least as great as its cost. The assumption in this case is that the user is the primary beneficiary. In an ideal situation, the costs should be shared when there are other beneficiaries of the service;

- *equity*: equal opportunity for transportation users is the second reason for requiring users to pay their way and for infrastructure builders to calculate these costs. The present methodology reflects equity in sense of equitable treatment; distribution and covering of costs;
- *finance*: in order to achieve the goal of 'user pays' it is necessary to calculate all the transportation and infrastructure construction costs. For those costs that cannot be fully understood, namely those not measured in monetary terms, corresponding estimation methods are elaborated trying to decrease the spectrum of errors and uncertainties. Although it can be difficult to measure these costs in the ideal case, they cannot be ignored.

Operation Flow of the Model

- Composition of ground data trends.
- Determination of cash flow sums.
- Calculation of usage and charges of sources.
- Preparation of output data tables.
- Calculation of financial indexes, drawing diagrams.
- Performing sensibility examinations.
- Generating new project alternatives which meet the threshold conditions, if needed.

Applications of the Model

After the above-described conception developed, the model using flexible limits of concerned effects is capable to carry out complex financial calculations extended for economical and social assessments. Among these in connection with transport infrastructure developments examinations concerned the M7 motorway along the fifth transport corridor, and the BILC investments projects have to be emphasised. As an example, Table 7.5 shows a simplified part of an output data sheet of the model calculated for the feasibility study of the BILC.

In the years following 2014 an internal rate of return of 9 per cent was reached which makes the infrastructure project feasible both for private investors and the state. A good cooperation between public and private is assumed at the BILC complex.

Table 7.5   A section of the monetary output table of the model (monetary values in 1000 HUF)

Macroeconomics	Year 2014
HUF CPI, %	3
CPI multiplier	2,135.063
EUR LIBOR, %	4
EUR CPI, %	3
EUR CPI index	1,589.127
HUF/EUR, Ft	312.704
Investment costs (1998), m. Ft	
Incomes at 1998 prices	
Expected income, containers	107.600
Expected income, RoLa	3.838
Income with BILC	5.066
Income without BILC	2.261
Income increase	2.805
Expected costs and savings	
Costs from increased traffic	1,542.75
Costs of personal	135
Personal cost saving in Józsefváros	105
Maintenance savings	100
Cost increase	1,472.75
Cost increase without savings	1,677.75
Cash flow	
Net cash flow	1,332.25
IRR	12.67%
EBRD debt	
Call	
Not called	0
Average not called debt	0.4
Fee for debt disposal	
Capital instalment	0.8
C. inst. flow, HUF	2,501.632
NPV of C. instalment	117.169
Interest paying	0.02
Int. p. flow	625.408
NPV of Int. p.	2,929.225
Cash flow with debt	
Debt flow	1,200.982
Cash flow after debt recovery	1,007.152
Internal rate of return	4.16%

Summary

This chapter aimed to give an overview of the developments established along the fifth corridor in Hungary recently. The establishment of up-to-date infrastructure is the only possible way for the country to improve its competitiveness, and to yield a profit from the transport sector. In order to allocate the capital to adequate projects, and to help the financial feasibility with favourable debts it is necessary to support financial calculation with an appropriate tool that can consider all necessary data from the special circumstances. This was and is the task of the Department of Transport Economics in cooperation with more private and public enterprises. There is still a lot to learn in this field, but the first steps have already been taken.

References

Berger, L. (2000), 'Development of Branches on Corridor V', in *Proceedings* of the conference on Europe's and Asia's Connection with the Fifth Pan-European Corridor, Zahony.

Putsay, G. (2000), 'Fast Speed Motorway-programme', *Magyar Nemzet* (*Road Construction*), Budapest (in Hungarian).

Tánczos, K. (2000), 'Euro-compatible transport Infrastructure – Requirements and Possibilities. Hungary at the Millennium – Transport, Communication, Informatics', *Transport Systems and Infrastructures*, Strategic Research in Hungarian Scientific Academy, Budapest, pp. 73–89 (in Hungarian).

Tánczos, K. (2000), 'Current transport Technology Development In European Union Countries. Studies – Transport', *Transport Systems and Infrastructures*, Strategic Research in Hungarian Scientific Academy, Budapest, pp. 9–37 (in Hungarian).

Tánczos, K. (2001), 'Basic Investigation Methods for Efficient Allocation of Sources in the Development, Operation and Maintenance of Transport Infrastructure Network', *Scientific Review of Communications*, 2001/9 (in Hungarian).

Tanczos, K. and Bekefi, Z. (1999), 'INNOFINance – Support Instrument for Financial Planning and Implementation of Large Scale Projects', *INNOFINance User Guide*, Budapest.

Tánczos, K., Duma, L. and Rónai, P. (2001), 'External Costs and Benefits of Waterborne Freight Transport in Europe', in *Proceedings* of the conference on European Inland Waterway Navigation, 13–15 June, Budapest.

Tánczos, K., Murányi, M., Orosz, C. and Gedeon, A. (1998/99), 'Establishing and Financing Large Scale Transport Investments in International Practice – Hungarian Useful Consequences', *Közlekedéstudományi Szemle*, pp. 332–40 (in Hungarian).

PART II
THE 'MISSING TRANSPORT LINKS' IN SOUTHEASTERN EUROPE

Chapter 8

Construction of a New Combined Road/Rail Bridge over the Danube at Vidin – Calafat and its European Integration Potential

Simeon Evtimov

Substantial bottlenecks exist along the international transport corridors, which impede major movement of people and goods to/from Southeastern Europe. One of most significant is the missing fixed link along Pan-European Transport Corridor IV over Danube River between Bulgaria and Romania. Development of transport infrastructure is a priority of the Bulgarian Government to provide strategic land connection alternatives between Europe and the Middle East. The proposed new link would foster mutual opening of countries to each other and serve the local, regional and the long-distance traffic in support of market orientation of economies. Selection of the bridge location at the crossing-point of the land-based and waterway corridors creates a good range of new transport logistics solutions with effect on the availability, condition, speed and loading capacity. The decision for construction of a new bridge at Vidin – Calafat is the response to real transport demand in Europe and will promote the integration process on the continent. The new bridge is an important policy instrument for regional stability and attracts international financing through the Stability Pact. The design standards for the project are based not on traffic forecast only, but also on requirements for a sustainable development of the region.

Project Background – Political and Economic Change in Europe

The transport and logistics infrastructure in Eastern Europe and the Balkans has undergone immense change over the last ten years. Railway systems have deteriorated and existing highway infrastructure has come under considerable pressure with rapidly increasing car ownership and use of freight vehicles on the roads. The events accompanying processes of disintegration of former Yugoslavia and the Kosovo crisis in particular, which interrupted at one and the same time River Danube navigation and the land route corridor via Belgrade, demonstrated that the lack of alternative routes for international traffic might break for a long period the normal trade links of an entire region. International humanitarian and peacekeeping forces' mobility during Balkan assignments was seriously hampered with the necessary increase of their routes by more than 250km and delays measured sometimes in days. Such a situation in particular provided the opportunity to realise that substantial bottlenecks existed along the international transport corridors, which impeded major movement of people and goods to/from Southeastern Europe. The most significant of them is the lack of adequate bridge capacity over the Danube river between Bulgaria and Romania (see Figure 8.1).

It is clear that such a picture is the result of the long-term policy of previous political regimes for closed borders and isolation from the Western world. Decades would be needed to overcome the disadvantages of underdeveloped transport infrastructure connections, but this process has already started.

Figure 8.1   Major transport corridors across Bulgaria

In Bulgaria the railway line to the western border with Yugoslavia has been electrified; rehabilitation and electrification of the Dupnitsa – Kulata rail line to Greece is carried out; construction of 'Struma' Motorway is planned in the same direction. Design work has been started for the upgrading of Plovdiv – Svilengrad rail line to the Turkish/Greek border, and construction works are to follow soon. Financing is provided for 'Trakia' and 'Maritsa' Motorways construction which, jointly with the Burgas Port reconstruction, will guarantee extension of Corridor IV to the Trans-Caucasian Area. In Romania also large-scale projects are implemented for modernisation of international corridors infrastructure for which more that €500 m have been allocated. Major reconstruction of the rail line Craiova – Filasi – Targu Jiu – Petrosani along the River Jiu valley at this phase would provide opportunity for alternative routes along international corridors, instead of construction of a new line via Maglavit to Turnu Severin. Electrification of the 107km railway line between Craiova and Calafat is envisaged too. In the light of all these measures for a phased improvement of the transport infrastructure, construction of the new Vidin – Calafat bridge will be considered one of the most significant steps to opening up Europe and attracting international transit traffic flows.

Project Rationale – Geopolitical Situation

There will always exist the probability and potential danger of interrupting international transport corridors, and such technical obstacles could arise not only as a result of damage to bridges, but also as a result of shipwrecks, accidents and failures in canal lock facilities, or even due to lowering of navigable river water levels. That is why the decision to construct a second River Danube bridge between Bulgaria and Romania was only accelerated by the situation around Yugoslavia, but in fact it was the logical result of broader and broader contacts between East and West parts of Europe, opening-up of the countries to each other, and the demand for development of alternative transport infrastructure in support of the growing trade links. The availability of a single fixed link for the land routes, in addition to the lack of any alternative for an entire region, did not correspond to the contemporary vision for market orientation and providing customers with the opportunity for choice. In this sense it is interesting to mention that Bulgaria and Romania are the only countries bordering the Danube which have no alternative bridge links over the river. Between Yugoslavia and Romania three bridges are in operation; between Hungary and Slovakia also three bridges are available, while between the two countries under discussion, only one bridge connection exists, at Rousse/Giurgiu. The closest neighbouring bridge crossings are located at the Iron Gate, 375km upstream, and at Cherna voda, 188km downstream from Rousse/Giurgiu. At the end of twentieth century such a situation may be even regarded as extraordinary, having in mind that even 2,000 years ago the Romans considered it quite normal to build bridges over the Danube in this area, such as the Turnu Severin Bridge or the Escus Bridge (at the Gigen/Corabia location). Why then today is it not considered normal that two neighbour countries, with a more than 400km common border along the Danube, should have not only one but two or even more bridge links, so that the river could be a connection between them instead of a line of separation, and more opportunities be provided for good transport links between the two countries, and between them and their neighbours? Activities for cleaning and opening the Danube for navigation and for construction of the bridge are elements of one and the same policy for ensuring regional stability and development, which is being carried out by the Stability Pact. Promotion of transport connections between the Southeastern European countries and the rest of the continent, including elimination of bottlenecks in the existing infrastructure, would stimulate economic contacts and trade, would push forward the integration processes, and ultimately would contribute to the stability of the region as a whole. The important political conclusion was also made that missing links in the transport corridors' infrastructure could be a source of regional instability (see Table 8.1). So, the decision to construct a new bridge at Vidin – Calafat is a response to a real transport demand in Europe and will promote the integration process on the continent.

The importance of Bulgaria as a transit country is growing, as its transport corridors provide the strategic land connection between Europe and the Middle East. Development of transport infrastructure is one of the priorities of the Bulgarian government. Construction of a second combined rail and road bridge over the Danube between Bulgaria and Romania at Vidin – Calafat is the first priority project anticipated for attaining this strategic objective. It is not the bridge that makes the corridor, as Corridor IV exists and its importance for all-European integration has been duly appraised. The lack of a bridge between Bulgaria and Romania however, represents an infrastructure bottleneck in the corridor and an obstacle for the normal development of transport and for full utilisation of the corridor integration potential (see Table 8.2).

The unstable political situation in Yugoslavia, caused by the Kosovo crisis and the war conflict, presented once again the urgent necessity of construction of the new bridge, in order to eliminate the missing fixed link on the route

Table 8.1 **Insufficient bridge capacity between Bulgaria and Romania**

Border crossing	Bridge capacity	International transport corridors effect
Yugoslavia – Romania	3 bridges	
Hungary – Slovakia	3 bridges	
Bulgaria – Romania	1 bridge	Bottlenecks and missing links Broken transport connections Low economic activity

Table 8.2 **Fostering effect of the new Vidin – Calafat Danube river combined rail and road bridge between Bulgaria and Romania on the European integration process**

Bulgaria – factor of regional stability

Economic and foreign policy reforms	Market orientation and European Union and NATO integration
Territory	Strategic land connection between Europe and Middle East
Investment priorities	Development of international corridors transport infrastructure
Strategic objectives	Improvement of East-West Europe contacts Elimination of a missing fixed links Development of alternative transport infrastructure

and give complete configuration of Pan-European Transport Corridor IV. The project was included in the Quick Start List of Regional Infrastructure Projects of the Stability Pact. The proposed new transport link would serve the local, regional and the long-distance traffic between Western and Southeastern Europe along the route of Pan-European Transport Corridor IV and further to the Middle East. The Bulgarian government strongly believed that an early beginning and completion of the works would have a positive impact on all neighbouring countries (see Table 8.3).

Table 8.3   Selection of bridge location

Selection criteria	Location advantages
Convenient transport infrastructure alternatives	Replacement of low capacity ferryboat by a fixed link
	Improved combined transport opportunities at crossing-point of land and waterway corridors
	Availability of approach road and rail infrastructure
Long-term solution	Perspective land transport route between Europe and Middle East
	Link between Thessaloniki port and Danube river
	Large emerging markets accessibility
Low cost solution	Maximum use of existing infrastructure
	Most favourable geotechnical conditions
	Minimum environmental impact
Regional socioeconomic impact	Positive impact on all neighbour countries
	Stimulation of European regional integration
	Good range of new transport logistics solutions
	Transport corridors synergy

Project Location – Transport and Investment Policy

In view of the current situation in the region with the altered conditions and the need for convenient transport infrastructure alternatives to overcome the consequences of the conflicts in Yugoslavia, all previous traffic forecasts scenarios are being revised. Even after improvement of the political situation around Yugoslavia the new bridge will be an important policy instrument for economic integration and stability through creation of transport alternatives. Corridor IV appears to be the only land alternative for Greece to the remaining European Union member states and, not only at present, but in perspective too, will serve as the basic land transport route connecting Central and Western Europe with the Near and Middle East. The existing ferryboat line does not correspond to the modern requirements for combined transport as per the international standards and presents a serious restriction to the development and growth of international transit traffic along this most important Trans-European corridor. Selection of the bridge location was based on continuous studies carried out by Bulgaria and Romania and by a number of international consulting companies. Availability of developed approach infrastructure on both sides to the future bridge is one of the arguments of important weight in support of the Vidin – Calafat location. It would be sufficient to remember that even when the 'classic' route via Belgrade was open the daily traffic operated via Vidin – Calafat Ferry was reaching 90 rail wagons and 750 road vehicles. The railway line to Vidin was built in 1923 with parameters for 80km/h, and on the Romanian side Calafat was connected by rail much earlier – in 1895. In the 1980s the UN Economic Commission (ECE) included these railway routes in the Agreement on the most important international lines for combined transport (AGCT), while the E79 road is categorised as Class A in the European Agreement on Main International Traffic Arteries (AGR). The direct link between the Danube and Thessaloniki port provides the possibility for transport access by Greece and by the international transit traffic to the big European inner waterway, and across the river to the unlimited markets to the north in Romania, Bielorussia, Poland and further, to Scandinavia. Going back to the 1930s, agreements have been concluded with Greece for facilitation of transit traffic, but they could not be applied '... due to the lack of a bridge over the Danube'. Already for several years, jointly with Greece and with the assistance of the European Union, practical steps have been undertaken for improving the standards of road and rail sections along the corridor. Being located at the crossing-point of the land-based and the waterway transport corridors, the future road/rail bridge will provide a good range of new transport logistics solutions with effect on the availability, condition, speed and loading capacity, which will undoubtedly result in journey time advantages (see Table 8.4).

Table 8.4   Comparison of alternative distances and time journeys by rail

Alternative routes (by rail)	Distance (km)	Journey time (hours)
Sofia – Belgrade – Budapest	772	15:20
Sofia – Vidin (Crayova) – Budapest	1,019	18:50
Sofia – Vidin (Maglavit) – Budapest	897	16:20

The results of the various studies completed by now confirm that the construction of the bridge at this location, which will serve both road and rail traffic along the Corridor, is the only true and long-term solution of the challenges of the upcoming century. Its regional importance is recognised as a link between two neighbouring regions in Bulgaria and Romania, which have been given the status of economically underdeveloped and with a high level of unemployment by the European Commission. The results also confirmed that the social and political significance of the bridge is a satisfactory compensation of any anxieties about the short-term economic feasibility of the construction costs. Relevant data from the database of the *Study Report on the Traffic Forecasts for the Ten Pan-European Transport Corridors of Helsinki* (NEA, 1999) made it clear that the presence of a bridge over the Danube at Vidin – Calafat would attract, even without increase of traffic in the years to come, at least 1.6 million road vehicles and 0.9 million rail tons. According to most recent forecast (BCEOM-SYSTRA, 2001), by the year 2030 traffic volumes could reach the average daily levels of 7,000 road vehicles, and 60 trains (18 passenger + 42 freight). Even a passing glance over the map would show that certain monopoly exists over the routes of transport corridors IV, VII, and X. It comes out that without an alternative under specific circumstances, entire regions of Europe could be isolated from transport point of view. So the new project would allow restoration of lost economic links of the region and preventing of future transport cut-offs for any reasons. Today's world is a world of alternatives and free choice between competing options and that is why monopolies of any type are not tolerated. If only for the reason of expected reduction of infrastructure charges and transport fares as a result of competition between the transport corridors, it is worth building the new bridge. Under the all-European transport policy for industry and trade promotion, the end customer of the transport service will feel the tangible benefits from integration in the European Economic Area and only in this way will the policy of sustainable mobility be implemented. Development of transport corridors is supported by joint investments aimed at creating various transport networks to meet growing demand for the movement of people and goods and to ensure application of the following market principles:

- elimination of any restrictions to offer transport services;
- freedom for the transport companies to respond to market requirements;
- freedom of choice of the transport mode on behalf of the customer;
- protection of environment;
- optimisation of public expenditures when solving the transport problems.

It is not considered that traffic forecast only would justify the bridge design and construction. Such a decision must also take into consideration the bridge specific functions as a cross-border facility, the requirements for a seamless combined transport services promotion and the perspective for the Corridor IV development. According to the Study of PATINVEST (2000), the design standards for the project envisage dual two-lane carriageway and single-track electrified rail line (with the possibility of a future upgrading to a double-track line) on the new fixed link over the Danube. The sustainable development of the region, the markets and the transport facilitation will be supported by the logistics opportunities created through the systematic combination of the relative advantages of the railways effective for mass transport at medium and long distances, namely reliability, regularity, safety, environmental compatibility and low power consumption, with the relative advantages of the other transport modes in *combined* services.

Facilitating transport services between the Southeastern European countries and the rest of Europe will stimulate economic contacts and trade, push forward integration processes and, ultimately, contribute to the stability of the region as a whole. From the viewpoint of history there is no 'classic' route, however. For hundreds of years the traditional trade flows between Europe and Asia Minor have passed via the Balkans, developing route alternatives across the central part of the peninsula, to the north of the Danube, or to the south of the Rhodopes. Such was the necessity in the past, such is also the necessity now, because the Balkan Peninsula itself plays the role of a 'bridge', a zone of connection of three continents: Europe, Asia and Africa. Such a role requires balanced development of both land and water transport infrastructure, in order to facilitate transit flows movement via the region. In that sense the shortage of competitive land route alternatives is a disadvantage of regional scale. So the construction of the new Vidin – Calafat bridge is justified both by historic and by contemporary assumptions for maintaining the geostrategic balance in the Balkans. It is an integral component of a group of priority strategic projects in the transport sector of Bulgaria, ranked as:

1   *elimination of the missing links in the existing infrastructure* (also including a new railway link between Bulgaria and Macedonia), and as next priorities;
2   *upgrading of existing transport infrastructure to improve quality of transport services* (covering the projects for reconstruction of the railway line Plovdiv – Turkish/Greek border, electrification of the railway line Dupnitsa – Greek border, and completion of main motorways in Bulgaria following international east-west and north-south routes, which are already being implemented); and
3   *construction of important infrastructure facilitating intermodal operations* (under this category comes the reconstruction of the Sofia International Airport which is also under way, and the construction of a new intermodal terminal in Sofia).

Mutual influence of various projects and alternative scenarios accordingly are of essential importance, as good coordination of their implementation would generate traffic, intensify utilisation of resources available and capital invested and ultimately achieve multiplication effect through synergy of economic and non-economic benefits. Systematic strategic analysis of the alternative routes (distances, journey times, investment costs, operational costs, etc.) shows that the selected location in the short-term aspect allows minimum investments and fastest implementation, while in the medium- and long-term aspects, under the scenario of construction of new and upgrading the existing approach infrastructure, represents an equitable alternative of Corridor X, which does not close the perspective competitive development of the entire Corridor IV. Although this would not be the shortest route between the Balkans and Central Europe, as a result of the new bridge project Corridor IV would provide the shortest competitive alternative of the Belgrade route without increasing the number of border crossings. A comparison between the distances and journey times by rail from/to Sofia shows that after the bridge construction, even without any upgrading of existing infrastructure along the corridor, distances increase not more than 32 per cent, and journey times not more than 23 per cent. More interesting are the comparisons that could be made for the routes Central Europe – Turkey and Central Europe – Greece. Such comparisons, however, reflect only the traditional domestic or so called 'national' aspect in selecting a route (see Table 8.5).

Table 8.5   Comparison of distances to/from Greece and Turkey – national approach

Direction	Route	Distance	Difference (%)
Budapest – Istanbul	Belgrade – Sofia	1,397	
	Calafat/Vidin	1,559	(+12)
	Giurgiu/Russe	1,678	(+20)
Budapest – Thessaloniki	Belgrade – Skopje	1,104	
	Belgrade – Sofia	1,126	(+2)
	Calafat/Vidin – Sofia	1,288	(+17)
	Bucharest – Giurgiu – Sofia	1,736	(+57)

The international aspect of calculations in providing the opportunity for selection of competitive alternatives of the transport connections Europe

– the Balkans, requires estimation to be made concurrently to/from Turkey and to/from Greece (see Table 8.6).

Table 8.6   Comparison of distances to/from Greece and Turkey –
European approach

Basic direction Europe – the Balkans

Via Yugoslavia		Via Romania	
Route to/from Greece	*Distance*	*Route to/from Turkey*	*Distance*
via Bulgaria	*(km)*	*via Vidin/Calafat*	*(km)*
Budapest – Belgrade – Istanbul	1,397	Budapest – Vidin – Sofia –	
Sofia – Thessaloniki	354	Istanbul	1,559
Summary for E-B Direction	1,751	Sofia – Thessaloniki	354
		Summary for E-B Direction	1,913
Route to/from Greece	*Distance*	*Route to/from Turkey via*	*Distance*
via Macedonia	*(km)*	*Giurgiu/Russe*	*(km)*
Budapest – Belgrade – Istanbul	1,397	Budapest – Bucharest – Russe	
Nish – Skopje – Thessaloniki	506	– Istanbul	1,678
Summary for E-B Direction	1,903	Russe – Sofia – Thessaloniki	648
		Summary for E-B Direction	2,326

This is exactly the approach of forwarders and operators (of the customers respectively) under the real market situation. Only this approach would guarantee the free choice between competitive routes for trade links, minimising in total infrastructure investments and operational costs. In this way it is seen that construction of the new Vidin – Calafat bridge as a component of Corridor IV would create indeed a comparable alternative infrastructure allowing transport connections not to be broken under any circumstances and achieving one of the major objectives of the modern European transport policy: *sustainable mobility*. Maintaining the ecological balance would also be a priority task for the designers and constructors. According to Denev (2000), as any other construction site, building of the bridge would be related to environmental risks, which will be assessed in advance by a number of environmental impact assessment (EIA) studies currently under way. Global long-term developments and the site conditions could require special river training solutions for the stability of the foundations of the bridge. In this respect the EIA needs to specify conservancy requirements and likely development trends in the preferred ecological system for the next 100 years, also under consideration of climate change. It must be kept in mind however, that the potential environmental threats from the bridge would be hardly bigger than potential dangers from river shipping. In that sense it may be even expected that the new bridge project would eliminate some of the existing risks and would also contribute to the environmental conditions improvement through the opportunities established for combined transport development and transfer of larger volumes of traffic from road to rail. Construction of the Vidin – Calafat Bridge as an issue of common European importance has been supported by the positive reaction of the European Commission and the International Financing Institutions about the Project Finance. Based on the bilateral Agreement between Bulgaria and Romania, the Government of Bulgaria shall secure the financing for the preliminary design of the Site, the technical design and construction of the bridge itself, as well as of the adjoining infrastructure on the Bulgarian territory, while the Government of Romania shall secure the financing for the technical design and construction of the adjoining infrastructure on the Romanian territory. It was quite normal for the international community to financially support the efforts of both countries to contribute to the opening of the Balkans to Europe, to improve their accessibility to the international markets and to overcome the economic disintegration and underdevelopment. The project will provide the construction of a combined (road and rail) bridge over the Danube River between the towns of Vidin and Calafat.

Project Parameters – Preliminary Studies' Results

A set of pre-investment interconnected studies were carried out financed through French and German government grants. They covered the preliminary EIA, geotechnical and hydrological survey, and an economic study plus cost estimate for the project. The region for the intended design of the bridge crossing is situated in the northwestern part of the Moezian Plain within the range of the Vidin Lowland. It is at a distance of about 49km downstream the Danube from the mouth of the Timok River. The geological structure of the terrain is relatively simple and is categorised with degree VII of seismic intensity. Bridge clearance is a sensitive item from hydrological viewpoint, especially when considering future developments that are difficult to predict. Two alternative navigation solutions have been analysed for providing minimum 13 m and maximum 20m height. A stretch of 796km of the Danube was confirmed to be the most suitable location, selected by the two countries for the bridge construction. In order to assess the investments needed possibilities have been analysed in the studies performed, of the application of various structures: steel, concrete, trusses, box girders, cable stayed. Cross sections of combined (motorway and railway) bridges have been of interest only. The estimated investments are approximately €220 m. According to a BCEOM-SYSTRA report (2001) the economic internal rate of return of this project would vary between 9 per cent and 16 per cent, depending on the macro-economics scenarios and the type of bridge construction. Future tender procedure however, shall provide the opportunity for competitors to propose their own design solutions, without being restricted to the type, material and method of construction, in order to give the employer the possibility of selecting the best proposed design. The construction period is expected to be three years. The intention of the Bulgarian government is to ensure funding for the project from following sources:

- Bulgarian state budget funds for preliminary studies, EIA reports, design and construction (up to €18 m);
- European Union PHARE grant funds for Project Preparation Facility;
- European Investment Bank loan towards design and construction (up to €70 m);
- Grant aid from the European Commission through ISPA towards design and construction (up to €75 m);
- Stability Pact contribution for studies, design and construction. Up to now contributions have been attracted in the form of grants or soft loans from the French Development Agency (AFD), German Credit Institution for Reconstruction (KfW), and the US Trade and Development Agency (USTDA) at the amount of approximately €23 m.

Analysis of socioeconomic data makes it possible to highlight the assets of the northwest region of Bulgaria and the southwest region of Romania. These are the basis for strategies, aimed at accelerating the regional development process, taking profit of the 'window of opportunity' opened by the new bridge project. Both regions have the privileged situation at the crossroads of three main European transport corridors (IV, VII and IX), which give them a strategic position to form a major multimodal transport area (rail, road and water). The rail/road networks of both regions are fairly well developed but require rehabilitation due to insufficient maintenance. Upgrading works should also be carried out at the right time, otherwise the current poor state of some roads and only partial electrification of the railway lines could become serious obstacles to attraction of international traffic, promotion of Vidin – Calafat site as an important transport centre in future, and in turn the economic development of the regions.

In relation to the new Vidin – Calafat bridge the areas on both sides of the Danube are to be used in the near future for setting up all kinds of ancillary services needed by operators of road, rail, water transport and intermodal activities. Particular attention should be paid in order to avoid overlapping or duplication. A well coordinated distribution of services would maximise the efficiency of the investments made. In this respect, initiatives to develop free trade zones, new multimodal and trans-shipment equipment, logistical platforms etc., should be planned jointly between Vidin and Calafat Municipalities, as well as other institutional partners concerned at regional and national level in Bulgaria and Romania. Construction of the bridge will create an integration link between the two towns, thus ensuring a good complementarity between investments and activities on both sides of the Danube, thereby developing the cross-Danube exchanges and cooperative relations, giving the best chance to boost the economic development of the cross-border region as a whole.

The increase in traffic and commercial turnover that the regions of Vidin and Calafat will experience will result in a growing demand for such services as hotels, restaurants, shops and trades of various kinds, services to businesses. The majority of the firms in these sectors are small and medium-size enterprises (SME), which should be the major beneficiaries of new opportunities that will emerge quickly with the bridge construction start. Up to now, SME have not contributed a large proportion of the region's economy. Development of new local activities directly or indirectly linked to the bridge construction, and the prospects offered by the growing needs for SME in various sectors, should contribute to slowing down the migration of young people observed recently. The bridge construction will of course not bring to an end all the difficulties, which the two regions are facing, and will not solve by itself the problem of unemployment, but its impact on local and regional economic activity, undoubtedly, will be significant. It should give a boost to the local economy and create new job opportunities, while the regions should at the same time benefit from economic growth at national level. European Union pre-accession policies should contribute to the acceleration of growth both at national and regional level.

The infrastructure works will result in generation of jobs, transfer of manpower, purchase of construction materials and other inputs in the regions. To assess the value of purchasing power injected in the regional economy, it has been considered that construction works will be carried out by local staff, with estimated 70 per cent of the salaries to be injected in local economy commercial turnover, while supervision will be entrusted to international experts, with only 50 per cent of the salaries estimated to be put in local commercial activities. The conclusion is that an average of 980 jobs would be created over the construction period, and €57 m from the beginning of works up to the bridge opening would benefit the local and regional economy. Apart from the direct impact of the bridge construction, indirect effects would additionally develop, as a result of expenditures of initial beneficiaries, generating additional incomes for 'second round' beneficiaries, etc. These indirect effects cannot be precisely estimated, as they depend on demand and consumer behaviour of each 'round' beneficiaries, but they might well double, in volume, the direct impact.

For the time being and in the absence of a bridge, transport over the Danube between northwest Bulgaria and southeast Romania is ensured by ferries and ships. Opening the Vidin – Calafat bridge will offer an alternative to waterborne transport, which will result in important changes and the emergence of new trends in the traffic patterns between the two regions. In the future, the recent creation of the Euro-region associating Vidin, Calafat and Zaichar districts, with the longer term process of European integration involving not only Bulgaria and Romania, but also Yugoslavia, shall create many opportunities for new cross-border economic and social relations inducing or generating new traffics on the bridge. According to the assumptions considered under various scenarios, opening of the bridge would multiply the traffic of road vehicles by numbers between two and more than three, depending on the economic growth and the tariff offered, in comparison to the traffic generated by local development only – as simulated in the 'no bridge' scenarios. A dynamic GDP growth would mean a rapid development of the regional activity, with the corresponding increase to the commercial exchanges and the local traffic. The main results from the vehicle traffic forecasts for the period until 2030 show a daily traffic level of buses, minibuses and small trucks or vans multiplied by around eight for the most cautious scenario, to around 20 for the most optimistic scenario.

Initial reference traffic forecasts have been made, with a few important variants concerning the bridge road tariff policy, which will be a key parameter, as from the opening of the bridge and in the long term. The impact of potential European integration of Yugoslavia was also explored with its impact in terms of cost reductions on Corridor X routes which are the most important alternatives to Corridor IV routes crossing the bridge. Total bridge traffic flows in 2030 are in the range of 3,000 vehicles per day in average in the 'small growth' macro-economics cases, 5,000 to 5,500 vehicles per day in average in the 'moderate growth' cases, and 10,000 to 12,000 vehicles per day in average in the 'large growth' cases. The long-distance traffic component has a low sensitivity to the tariff offered, but the total traffic is very sensitive to the 'frequent user' tariff opportunity, because of the growing importance of the local traffic component that one can expect. Simulating the effect of the European integration of Yugoslavia and Corridor X competition shows quite a limited drop in the total bridge traffic, being smaller in scenarios with rapid economic growth – 5–6 per cent, compared to 9–10 per cent in the scenarios with low economic growth. The results are about the same in 'without bridge' and 'with bridge' scenarios. It has to be mentioned that in the 'without bridge' reference scenarios, all traffic figures for 2030 are well beyond the maximum capacity of the existing Vidin – Calafat ferry platforms (provisionally estimated in the order of 1,000 vehicles per 24 hours).

The construction of the Vidin – Calafat bridge would open a new railway route on Corridor IV, which by itself could impact the conditions of competition between the maritime and land routes for freight transport. This impact however would be significant only provided that railway operators implement dynamic commercial policies, allowing forwarders to take profit of the cheap transport costs potentially offered by the new railway route. Rail transport for all origin-destinations in the region offers a very large cost advantage compared to all other modes (road, maritime and road, and maritime and rail), except for origin destinations in Greece for which the maritime and rail mode is the cheapest. This means that future changes in the conditions of competition will depend on factors other than costs – time, flexibility, reliability, except perhaps for a few origin-destination couples (Greece-European Union and Central European countries).

And finally, a conclusion can be drawn that construction of a combined (road and rail) bridge at Vidin – Calafat is much more advantageous, compared to construction of a road only or rail only bridge, as in the first case the positive impact would be complex and more comprehensive both in local/regional, and in continental aspect, while in the second case major benefits would be derived respectively for local/regional traffic development, or for long-distance traffic growth.

References

BCEOM-SYSTRA (2001), 'Economic, Financial and Social Analyses for the Construction of a New Bridge over the Danube River (Vidin – Calafat)', Report on Phases 1 and 2, Ministry of Transport and Communications-AFD, Paris, France, June.

Denev, Dobrin (2000), 'Analysis of the Impact on the Environment of Bridge over Danube River between Vidin and Calafat,' ET 'IRIN – ID', Sofia, Bulgaria, October.

NEA-INRETS-IWW (1999), *Traffic Forecast on the Ten Pan-European Transport Corridors of Helsinki, PHARE Multi-Country Transport Programme*, Rijswijk, The Netherlands, December.

PATINVEST Engineering (2000), *Study for Elaboration of tentative Values on Preliminary Designs for the Construction of a Second Combined Rail and Road Bridge over the Danube River at Vidin – Calafat km 796.000 and the Access Roads to the Existing Transport Infrastructure*, Ministry of Transport and Communications, Sofia, Bulgaria, September-October.

Chapter 9

The Role of the Corinth Canal in the Development of the Southeastern European Short Sea Shipping

Evangelos Sambracos

This chapter aims at identifying the role of the Corinth Canal in the development of the short sea shipping transport system of Southeastern Europe. The Corinth Canal, construction of which was completed in the nineteenth century, is considered to be one of the biggest technical undertakings in Greece. It serves the sea transport of goods as well as human mobility on not only a national but also an international scale. The canal can operate as a node of the transport network of Greece and also of the short sea shipping system of the Southeastern Europe. This is due to the important advantage of the faster and safest sea route that the canal offers for the transportation between the ports of the Black Sea and Eastern Europe with those of the West Greece, the Ionian Sea and the greater region of Central Europe. In order for the Corinth Canal to adjust to the ongoing developments and trends that have to do with the promotion of environmentally friendly means of transport, the integration of the different modes of transport and the establishment of a single intermodal transport network, its management has to proceed to the development and materialisation of a suitable policy. The chapter includes the findings of a survey conducted for the Corinth Canal authorities and concludes with certain propositions so that the Canal can revitalise and upgrade its position in the short sea shipping network of the Southeastern Europe.

Introduction

The Corinth Canal and its contribution to the rational development of the short sea shipping network of the Southeastern Europe, is examined in this chapter. Greece itself shows a special geopolitical importance as it is situated at the crossroads of three continents – Europe, Asia and Africa – and can be therefore considered as a natural bridge between Europe, Middle East and North Africa. Additionally Greece can be considered as the southeast gate of the European Union.

The geographic location of the Corinth Canal serves the cohesion of the Greek port system as well as the wider area between the Eastern and Western Mediterranean Sea and also the Black Sea ports. The importance of the Canal for the short sea shipping network of the Southeastern European region lies mainly in the advantages that the Canal offers to the sea trade at a national and international level. These are examined through a research survey that took place in 1999 regarding the possible expansion of its market.

Through the general directions of the European Union on the development of short sea shipping, the chapter focuses on the role that the Canal can play in it.

The Development of Short Sea Shipping in Europe

According to the prevailing notion and as recommended by the European Union, short sea shipping refers to the movement of cargo and passengers by sea between ports situated in the European area or between European ports and ports in third countries that have a coastline and their sea is adjusted to the European Union. Short sea shipping includes domestic and international maritime transport, excluding ocean-crossing shipping, and includes the feeder services (the filling and emptying of the goods (mainly containers), the redirection from or towards an open sea service to one of these ports/hubs) along the coast to and from the islands, rivers and lakes. It therefore refers also to sea transport between the member states and Norway, Iceland and other members of the Baltic Sea as well as northeastern Europe including the Black Sea and the Mediterranean Sea.

According to the European Union the European Short Sea Shipping fleet constitutes 40 per cent of the world fleet, while at the same time, the open sea fleet is 50 per cent of the corresponding world fleet.

The evolution of the European Union sea trade is presented in Table 9.1 for domestic and international intra-European sea transport, according to which domestic goods transport has increased by almost 67 per cent while transport between member states has increased by almost 144 per cent.

Short sea shipping has been growing in the past decade in the European Union. Goods carried increased by 17 per cent between 1990 and 1997. Of the total tonne kms in the European Union, the shares of short sea shipping and

Table 9.1   Evolution of the European Union goods sea trade

	1970	1980	1990	1992	1994	1996	1997	1998
Sea domestic transport (in 000 m. tonne kms)	97.3	147.2	152.4	148.5	145.4	159.3	156.5	162.5
Sea international intra-European Union transport (in 000 m. tonne kms)	374.9	632.8	770.0	826.2	810.4	869.8	916.5	914.0

Source: Eurostat.

road are almost equal. In terms of international tonne kms, short sea shipping has by far the largest share.

Regarding the short sea shipping fleet and its composition on the basis of the main goods transported, at the beginning of the 1990s dry cargo vessels represented more than 50 per cent of the fleet, while the percentage for liquid cargo vessels was below 20 per cent. Greece was second in the number of vessels after the ex-Soviet Union fleet, but came first in the European Union Short Sea Shipping fleet with almost 12 per cent of the total short sea shipping European Union fleet. Additionally, European Union member countries represented a significant percentage of the total short sea shipping in Europe (almost 45 per cent). In the following years, according to available data (European Union, 1999) from the European Union, the fastest growing segment of short sea shipping from 1993 to 1997 has been the containerised cargo, which rose by 44 per cent (in tonnes).

As to the determination of the vessel types operating in the short sea shipping system there are no predetermined dimensional characteristics. On the contrary, the dimensions and the tonnage of the vessels vary according to the market they operate, the type of cargo they carry and the volume of the shipment. Many operators in the market of dry cargo consider the short sea shipping vessels to be up to 3,000 dwt (deadweight tonnage) while others regard it to be around 6,000 dwt but 10,000 dwt is regarded to be the maximum tonnage level in intra-European Union sea trade (Tinsley, 1991). In this context we see handy size vessels used for the carriage of grains from Britain and southern France to the Mediterranean, Panamax vessels with self-loading possibilities in Great Britain's coastal transport lines and even container ships up to 1,000 TEUs in intra-regional activities. The average gross tonnage for vessels operating in short sea shipping is determined at 1,654 grt (gross registered tonnage) in the European Union and there is a trend towards increasing it while the upper limit is determined around 6,000 grt (Peeters et al., 1995).

Greece and the Promotion of Short Sea Shipping in European Union

Short Sea Shipping in the European Union

It is expected that the development and establishment of a single market, the liberalisation of markets and the removal of all obstacles in trade will boost intra-European Union trade. Different perceptions of demand and the notion of just-in-time in production, and therefore in transportation, point out the importance of time and also quality of service. In this context efficient and effective transport connections are of vital importance.

Within the frame of free and unbiased choice of transport means, the promotion of short sea shipping is based on the supply of a sustainable and efficient alternative solution for the products carried and the transport units that can be transported by all means of transport. Sea transport has many advantages to offer to the European Union transport system since it contributes to the relief from pressured and congested road networks and the cohesion of the market, as well as to the revitalisation of regional ports and their hinterland.

The continuous increase of short sea shipping in Europe and the special focus of the European transport policy on sea-borne trade is mainly attributed to the advantages that short sea shipping shows. Within the context of sustainable mobility and development, there are three main reasons for promoting short sea shipping, and these include, as reported by European Union (European Union, 1999):

- the promotion of the general sustainability of transport. Short sea shipping is emphasised as an environmentally friendly and safe alternative, in particular, to congested road transport;
- the strengthening of the cohesion and the facilitation of connections between different states and between regions in Europe and the revitalisation of peripheral regions;
- the increase of the efficiency of transport in order to meet current and future demands arising from economic growth. For this purpose, short sea shipping should be developed into an integral part of the logistic transport chain and also a door-to-door service.

The above justify the increasing effort of the European Union to increase the use of short sea shipping as an alternative transport means within the European Union and against the externalities the transport system shows (congestion, pollution, accidents etc.) that burden sustainable development.

The Role of Greece in the Mediterranean Sea-borne Trade

In the sea-borne trade in the Mediterranean, Greece has developed into a node in the international transport network. More specifically, Greece plays a significant role in the international sea trade route in the Mediterranean Sea, from the Suez Canal to Gibraltar. In this network Greece, through its ports, has a double role to play (Sambracos, 1999).

Firstly, it is a point of destination, where open sea vessels carrying out international trade serve the import of goods covering domestic demand. In this case goods arrive at Greek ports (mainly in Piraeus and Thessaloniki) and are then transported to the mainland through the road network and to insular Greece through the domestic coastal shipping fleet. Secondly, it is a point of goods' trans-shipment and transit to other countries of Southeast Europe. The ports serve as hubs, where freight (usually containers) is unloaded from mother ships, consolidated and redistributed to other countries with small vessels forming a short sea shipping–feeder network or by using the land transport network (road, rail) to the Balkans peninsula and from there to the rest of Europe.

Piraeus and Thessaloniki have a strategic position in the Greek port system, serving both the import of goods and their trans-shipment to other neighbouring countries. Available data on both ports show that the total freight traffic in Piraeus has increased 48 per cent over the period 1994–98 and 14.5 per cent in the Port of Thessaloniki between 1997 and 2000. Container traffic has shown an increase of 80 per cent in Piraeus and of 37 per cent in Thessaloniki.

The Corinth Canal in the Greek Sea Trade System

General Characteristics and Data

The geographical position of the Corinth Canal serves the cohesion of the Greek port system since it connects western with eastern Greece (Sambracos et al., 2000). Its importance as a link between western and eastern Greece was recognised early, and its construction was finalised at the end of the nineteenth century. Its dimensions reflect the sea transport market of that period, the vessels and the trade that was then conducted (Table 9.2). Taking into consideration the data presented in Table 9.3 it is concluded that the average tonnage is determined at around 425 nrt (net registered tonnage).

Table 9.2   Dimensional characteristics of vessels passing the Canal

Vessels' width (m)	Max. draught (m)
16.0–16.6	6.2
15.5–16.0	6.2
15.0–15.5	6.6
14.5–15.0	6.8
14. –14.5	7.0

Notes

1   The max. draught for vessels with width smaller than 14m is 7.2m.
2   The max. width of a vessel to pass the Canal is 18.3m.

Table 9.3   Traffic through the Canal

	No of transits	Tonnage nrt	Average tonnage nrt
1990	10,109	4,418,850	437.12
1991	9,639	3,970,237	411.89
1992	10,653	4,435,122	416.33
1993	11,018	3,799,754	344.87
1994	11,853	4,462,668	376.50
1995	12,545	5,625,123	448.40
1996	12,459	5,748,401	461.39
1997	11,026	4,950,959	449.03
1998	10,662	4,502,325	422.28
1999	11,011	4,897,925	444.82
2000	11,715	5,464,824	466.48

Source: Corinth Canal S.A.

As for the types of commercial vessels they can be categorised into the following main categories:

* freighters that carry bulk and general cargo;
* tankers, LPGs;
* ro-ro, container vessels;
* passenger ships, ferryboats, professional tourist vessels (carrying over 25 passengers);

- sailing boats and yachts (private and professional), professional tourist vessels (carrying fewer than 25 passengers);
- other vessels (diesel vessels, military ships, Port Police, tugs, fire vessels, lifeboats etc.).

For cargo carrying vessels the average tonnage is around 650 nrt, for tankers it is 458 nrt and for passenger vessels it is 1583 nrt.

For 1998, the composition of the Canal's traffic (in number of transits) is presented in the following figure, according to which freighters and tankers are the most regular customers since they have performed the majority of transits, followed by the private sailing boats and yachts respectively (Figure 9.1). For the same year, the majority of the vessels (43 per cent of transits) were under the Greek flag, operating on domestic routes, while 40 per cent of transits referred to vessels under foreign flags, operating between the Black Sea and eastern Mediterranean Sea to Greece or calling at ports in Albania and Yugoslavia up to Italy. A small percentage (4 per cent) were vessels under foreign flags operating between South Africa–Gibraltar, or vessels under the Greek flag operating between the Mediterranean Sea and north Europe to Greece. The rest of the vessels were barges, floating cranes and dredgers, as well as Greek professional tourist and fishing boats registered at local registries.

Additionally, for the years 1980–97, and for the vessels over 100 grt and under the Greek flag, an average percentage of 46 per cent and 45 per cent of the total transits referred to freighters and tankers respectively, while passenger vessels' share was only 6 per cent. In terms of tonnage (in nrt) freighters have by far the biggest share (48 per cent on average), followed by tankers (33 per cent on average) and passenger-carrying vessels (17 per cent on average).

The Importance of the Corinth Canal for Short Sea Shipping

Advantages for Navigation

The importance of the Canal for navigation is great, considering that the Peloponnese round is avoided, with positive consequences for travel distances, time and safety for national and international trade. In particular, the short sea shipping navigation between Black Sea – Aegean Sea – Ionian Sea – Adriatic Sea has to chose between circumnavigating the Peloponnese or passage through the Canal. At the same time, the pivotal role of the port of Piraeus, which is situated on this route, plays a significant role in the operation and promotion of

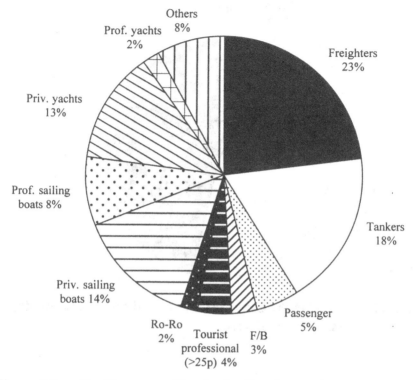

Figure 9.1 Traffic composition in the Canal (transits)

the short sea shipping concept if there are advantages in favour of the Canal.

The advantage of the Canal in comparison with the route around the Peloponnese is summarised in Table 9.4, which presents the travelling distances (in nautical miles) of the alternative navigation routes between the main destinations served by the Canal.

From the above data, it can be concluded that:

- the Canal shows great advantage for the domestic navigation between the Piraeus port and the ports in the Gulf of Patras;
- the advantage for sea trade between Italian ports and the port of Piraeus as well as between Ionian ports and Piraeus is also significant;
- the advantage between the ports of Adriatic sea and Ionian sea with the ports in North-East Aegean and Black Sea is satisfactory;
- finally, the advantage between the ports of the west and southwest Mediterranean and the north and northeast Mediterranean is minimal.

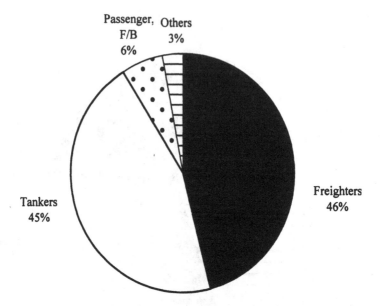

Figure 9.2   Traffic in the Canal 1980–97 (no. of transits, Greek flag vessels over 100 grt)

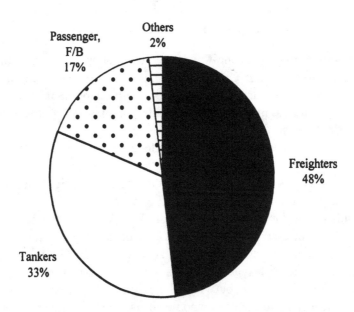

Figure 9.3   Traffic in the Canal 1980–97 (tonnage in nrt, Greek flag vessels over 100 grt)

Table 9.4   Advantage of the Corinth Canal in travelling distances

| Origin | Destination | Distance (nautical miles) | | |
		Through the Canal	Through the Peloponnese	Difference
Messina	Piraeus	403	477	74
Venice	Piraeus	721	851	130
Brindisi	Piraeus	333	464	131
Corfu	Piraeus	237	370	133
Patras	Piraeus	100	295	195
Messina	Sounio	428	463	35
Venice	Sounio	745	837	92
Brindisi	Sounio	358	450	92
Corfu	Sounio	262	355	93

Source: Ministry of Merchant Marine

The reduction in distances of about 100 nautical miles average also means time economy for the user of the Canal, which can be translated into further economy in fuels and lubricants and seamen's working hours as well as reduction in damages. All these result in lower variable operating costs. Additionally, the circular Peloponnese route involves higher navigation risk and danger, especially during the wintertime. The risk is higher for non-loaded vessels of up to 1000 grt. The contribution of the Canal to navigation safety means fewer accidents for transit vessels, reducing the social cost of transport. Finally, trading is a function of distance and time. The Canal, by reducing distance and travelling time, has resulted in greater demand for freight transport and has promoted the development of new trade zones.

The Importance of the Canal through a Research Survey

The importance of the Canal for the national and also international trade can be seen through the results of a survey that took place during summer 1999 on account of the Corinth Canal SA and in cooperation with the University of Piraeus (University of Piraeus, 1999). The survey was conducted through two subsurveys that took place in parallel as following:

- the first survey referred to vessels that went through the Canal between 28 June and 23 July 1999. The survey took place in the Canal and

focused on three main categories: freighters (bulkers, tankers, ro-ro, containers), passenger ships (including ferryboats), and leisure boats. The categorisation was based on the pricing policy of the Canal, which includes six pricing categories according to the type of the vessels and their origin/destination;
- the second survey referred to the maritime companies that were or still are customers of the Canal.

The surveys were conducted through questionnaires which included questions regarding:

- the vessels and their routes;
- the use of the canal and the factors affecting the demand for transport services;
- the customer's opinions regarding the quality of the services;
- the possible additional services the passengers might require.

In this period 680 vessels crossed the Canal and from these 188 questionnaires were gathered. The sample (%) of the vessels questioned in comparison with the total vessels that crossed the canal per type was:

- 17 per cent of bulk carriers;
- 25 per cent of tankers;
- 45 per cent of ro-ro vessels;
- 20 per cent of passengers;
- 50 per cent of ferryboats;
- 24 per cent of leisure boats.

Questionnaires were also completed by 35 maritime companies who owned a fleet of 228 vessels using the Canal's services. We should mention here two facts: firstly the majority of the Canal's companies/customers referred to the same operator, who was the one to answer on behalf of the companies he operated; and secondly, some companies were not the owners of the vessels but agents, who were unable to express the owners' opinion regarding the Canal's services.

The results of the survey are presented for the cargo vessels in the following paragraphs.

Freight Carriers

Among the freight carriers who took part in the survey 51 per cent were bulk carriers, 30 per cent were tankers and 9 per cent were ro-ro vessels. The majority of the vessels (55 per cent) were under the Greek flag while the rest were under the flags of the Ukraine, Russia, Malta, Albania, Panama and Germany. The tonnage of the vessels was up to 2,000 nrt, while the majority of vessels were between 100–1,000 nrt (21 per cent were above 1,000 nrt). As for their employment, over 60 per cent of the vessels were tramp vessels, while the majority of vessels under the Greek flag (especially tankers) were liners with regular and frequent transits through the Canal.

The majority of the vessels (64 per cent) reported that they used the Canal on a monthly basis while 18 per cent used the Canal only for 2–7 transits per year. The transits per year are presented in Figure 9.4.

Regarding the use of the Canal's service and the circular Peloponnese route, 78 per cent of the vessels responded. Of these, 28 per cent reported that in certain cases they have chosen the latter. Most of them made regular transits through the Canal and almost half of them were over 1,000 nrt. The main reasons for not using the Canal were related with the cost of the transit,

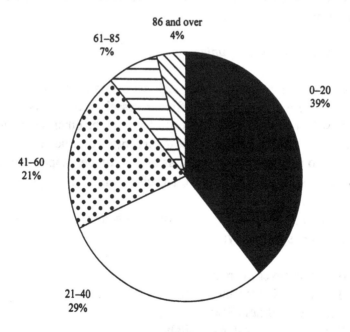

Figure 9.4   Regularity of yearly transits

time constraints and the favourable weather conditions. Most of them were tramp vessels.

As for the reasons for using the Canal, the determining factors for liners were:

- time economy was reported to be the most important reason for using the Canal;
- fuel economy (90 per cent of vessels);
- risk and danger avoidance (80 per cent of vessels);
- reduction of damages and wear and tear (70 per cent of vessels);
- lack of anchorages (60 per cent of vessels).

For tramp vessels the determining factors were:

- time economy was reported to be the most important reason by 88 per cent of vessels;
- avoidance of risk and danger (53 per cent of the vessels);
- fuel economy (47 per cent of vessels);
- reduction of damages and wear and tear (35 per cent of vessels);
- lack of anchorages during the circumnavigation of the Peloponnese (30 per cent of vessels).

Survey of the Maritime Companies

A total 35 maritime companies with 228 vessels participated in the survey. The majority were companies with freight vessels and tankers (22 and six companies with 140 and 20 vessels respectively). The other companies have ro-ro vessels, containers, LPG, etc. The main cargoes transported were general and bulk cargo (dry and liquid) such as grains, timber, paper, ore, cement, fertilisers, salt, wine, oils, etc., as well as chemical products, diesel, oil, gas oil, olive oil, etc.

The main routes followed by each category were in the area of southeast Europe and more specifically:

1 Cargo vessels cover the routes:
- Black Sea–Mediterranean Sea;
- Black Sea–Adriatic Sea;
- Black Sea–Turkey–Greece–Italy;
- Greece–Italy;

- Greece–Albania;
- East Mediterranean–Italy–Yugoslavia;
- Scandinavia–Greece–Cyprus.

2 Tankers cover the routes:
- Athens–Rio;
- E. Mediterranean–Black Sea;
- Mediterranean–Black Sea;
- Israel–Italy,

3 Ro-ro vessels cover the routes;
- Greece–Italy;
- W. Mediterranean–Adriatic–Greece–East Mediterranean;
- Black Sea–Adriatic Sea;
- Piraeus–Adriatic Sea;
- Slovenia–Piraeus.

4 Container ships usually operate in East Mediterranean and LPGs in the route Greece–Italy.

As for the reason for using the Canal, the majority of the companies reported as main reason time economy, followed by economy in fuels and lubricants. A third reason was the dangerousness and risk of the Peloponnese route, especially in winter. Some of them also cited wear and tear on the vessels and the lack of anchorages during the journey. Among the companies that used the Canal, some of them reported that they also use the Peloponnese route, mainly because of the cost of the Canal's transit and the possible delays. Thus where the difference in time between the two alternatives is small, then the Peloponnese route is preferred (for example, companies prefer the Canal when the port of origin is in the south Adriatic Sea and the port of destination on the north Aegean or Black Sea).

The Canal's Operation from the Demand Size – Conclusions

As already presented above, the Canal mainly serves small and medium-sized vessels with a tonnage of 500–2,000 nrt. The reasons for using the Canal are, in order of importance:

1 time economy;
2 fuel economy;
3 safety;
4 reduction of wear and tear and damage;

5   lack of anchorages on the circular route.

Both liners and tramp vessels gave the above factors the same weight of importance. These factors seem to overwhelm the factor of 'transit cost', since most of the commercial vessels considered it to be the main reason, along with delays (average waiting time 30 min.–1 hour), for not preferring the Canal. In spite of this, vessels usually used the Canal for its advantages and only preferred the circular route when there are no time constraints and the weather was good.

The survey highlighted the adequacy of the currently offered services (piloting and towage), but pointed out that the way the payment is conducted causes delays to the vessels and that it is essential to develop intermediate ways of payment (through bank accounts) or electronic collection systems, especially for regular users. Additionally, users proposed the development of new services. For cargo vessels, the most important new service was the provision of fuel and foodstuff, as well as the possibility of embarkation and disembarkation in the area of the Canal.

It is therefore understood that the Canal plays an important role not only for the Greek fleet and the Greek goods transport but also for the short ship shipping industry of Southeast Europe. We can confirm that most of the users of the Canal undertake regular transits. If there was no interest, the Canal and the specific market would surely already have been abandoned.

Discovering Measures for the Upgrading of the Canal's Role

The position and the dimensional characteristics of the Canal (small and medium-sized vessels) already serve the main routes in the Southeast Europe short sea shipping trade. As indicated from the survey, the Canal could become an important node in the national and international short sea shipping market, since it allows for the cohesion of the east and west Mediterranean Sea.

In order for the Canal to exploit the current developments and trends regarding the promotion of sea transport, its management needs to proceed to the development of a suitable operation policy. Such a policy should focus on both macro- and micro-economic short-term objectives.

In this context, the Canal's orientation should be towards three main policies: pricing, investment and institutional framework.

The Pricing Policy of the Canal

The price of the offered service is one of the major factors determining the level of demand, especially in those niches of sea transport that show high elasticity of demand. So far the structure of the current pricing policy is determined according to the following pricing principles:

• all vessels wishing to transit the Canal pay fees (towing and piloting). The payment of tolls and other dues varies depending on six categories which have been developed based on the following factors:
  – type of vessel;
  – origin and destination of vessel;
  – vessel's flag;
  – vessel's dimensional characteristics (tonnage, length, etc.);
  – type of some floating buildings.
  These categories include the following:
  – *Category A*: passenger vessels, freighters or motorboats sailing under the Greek flag (except barges, sailing boats or floating equipment), operating solely on domestic routes or making occasional calls to foreign ports;
  – *Category B*: passenger vessels, freighters or motorboats which have called at ports in Albania, Yugoslavia and then in Italy, as far as Taranto. Also, vessels which come from the Black Sea and Eastern Mediterranean Sea (as far as the port of Alexandria) heading to a Greek port and vice versa. Vessels under a foreign flag that have their origin or destination in Greek ports;
  – *Category C*: passenger vessels, freighters or motorboats which have a port of origin or destination in Southern Africa (further west than Alexandria) as far as Gibraltar. Vessels under the Greek flag that have as port of origin or destination in the western Mediterranean Sea, northern Europe and call at ports in western Greece or the Corinthian Gulf;
  – *Category D*: barges, floating cranes, dredgers, etc.;
  – *Category E*: Greek professional tourist and fishing boats under 100 nrt registered in local registries (Isthmia, Loutraki, Corinth);
  – *Category F*: sailing boats and yachts (excluding professional tourist vessels carrying over 25 passengers).
  Special categories should allow for increases in the fees (for all shipwrecks, night passages, passages during holidays, use of more than one tow or

pilot, etc.) or reductions (for professional tourist vessels carrying less than 25 passengers).

The applied pricing policy favours vessels sailing under the Greek flag operating mainly in the domestic market. Conversely, the fees are higher in the case of the vessels of category B, mainly due to the fact that these vessels enjoy many advantages when using the Canal in comparison with other vessels. Therefore, there is a price discrimination philosophy based on the level of benefit that the user enjoys when passing through the Canal and on their flag. There is no connection whatsoever between the fees and the cost of the Canal's services (tolls, pilot, towage).

According to the survey on the vessels and the companies, we can see that 33 per cent of the companies consider the transit fees to be reasonable, while the remaining 66 per cent of the companies consider the fees to be expensive. Accordingly, from the survey, the majority of them consider the fees to be reasonable.

Taking into consideration the results of the surveys and the general pricing policy principles, the implementation of a price discrimination policy (Sambracos, 2001) according to the user of the canal and the development of special prices should be the base of the pricing policy that needs to be applied. The proposed policy is based on the following key aspects:

- abolition of flag discrimination;
- categorisation of the users who operate on domestic and international lines;
- pricing based on free competition criteria;
- determination of the elasticity of the users.

For national lines, the pricing policy should focus on the marginal or average cost which would also maximise the economic welfare. For international lines, pricing should be according to what the market can bear, as determined by the elasticity of the user and the differential cost between the Canal and the Peloponnese circumnavigation. Additionally, a form of user discrimination policy should be included, in favour of regular customers or customers who schedule their transits.

The Investment Policy

The existence of suitable infrastructure in the Canal area is an additional factor that determines the demand for transport services. This infrastructure includes technical issues such as berths and breakwaters, as well as towboats to serve the passage and temporary anchoring of the vessels.

In the long term, such policy should examine the possibility of enlarging the Canal's dimensions so as to support the passage of bigger vessels, based on the results of a feasibility study. In the short term, the development of suitable facilities and equipment is important so as to facilitate the requirement for continuous communication and information diffusion. Information technology is a supportive tool that acts as an enabler covering the need for communication in the Canal and also between the Canal and its users. Thus the development of new facilities (banking, supplies, etc.) should also be considered.

The Institutional Operating Framework

According to the existing institutional operating framework, the Canal is a public organisation whose status does not allow for any commercial initiatives apart from the transit of vessels. This restricted institutional framework needs to evolve, bearing in mind the demand and the developments towards a more integrated transport system. The Canal is a node in this network and therefore should be able to offer new services of a largely qualitative character. They include new investment plans in accordance with the market demand, such as:

- additional port facilities, berths, marinas, piers, etc., for the temporary anchoring of commercial vessels in order to take on or disembark seamen;
- additional services which could include the procuring of fuel, lubricants, water, spare parts, etc.;
- sufficient connection with the other modes of transport (intermodality);
- reception facilities, including restaurants, rest areas, banking services, etc.

Regarding the institutional framework, it should be mentioned that in recent years the proprietary regime of the Canal changed, and at the time of writing the operation of the Canal (not the infrastructure) was undertaken by a private company.

Conclusions – Epilogue

The Canal and its dimensional characteristics allow for its further development within the short sea shipping trade market. In particular, the Canal serves the operation and development of feeder services, since it allows for the transit of small and medium-sized vessels up to 2,000 nrt covering short distances between the eastern and western Mediterranean and also Black Sea ports.

Knowing the objectives of the European Union regarding sustainable transport of goods and minimisation of transport externalities, there should be a special focus on the Canal for the service of the southern eastern European edge. The existing congested road networks on one hand, and the lack of efficient infrastructure in many countries (especially those not belonging to the European Union) on the other hand, upgrades the role of sea transport in this area. This is due to the fact that sea is the most economic means of transport and there is no demand for infrastructure since it possesses natural navigable corridors. The incorporation of sea transport, and more specifically short sea shipping, into the intermodal transport chain is also a perspective seriously observed by the European Union.

Another aspect is the connection of short sea shipping with inland navigation in Central and Eastern Europe and through the Canal. Our region offers the possibility of the creation of a new navigable network connecting the Mediterranean Sea with the Danube. The importance for the Balkan countries of such an undertaking can be appreciated through several arguments, such as:

- access to Europe's navigable network;
- access to new, rapidly-developing markets of Central and Eastern Europe;
- development of a new land transport system and entrance/exit of these countries to the Mediterranean Sea;
- further development of the competitiveness of this means of transport resulting in the reduction of the final price of their products in the European markets;
- development of the sea and river ports.

Finally, in considering the Corinth Canal in connection with the two main ports of Greece, Piraeus and Thessaloniki there are a number of important points to be made.

Firstly, the fact that Piraeus is close, in terms of distance and time, to the main commercial routes of the Mediterranean provides it with the opportunity to become a hub port for Southeast Europe. This could be achieved in one of two ways (Sambracos, 1988, 1999). The first includes the open sea–road/rail transport. This means that the goods would be unloaded by mother-ships in the port of Piraeus and then carried to the other Balkan countries and Central Europe through the existing road/rail networks. The second one includes sea–river transport. The goods would be trans-shipped in Piraeus or Thessaloniki and would then be carried to the other countries through the Danube and also to the West Mediterranean Sea by feeder ships via the Aegean and Black Seas. We should also consider the possibility of a short sea shipping–inland navigation network from Thessaloniki via Axios–Vardar river to the Danube (Sambracos, 1999).

The second alternative is the most appealing one since the road or rail network would not be used. In this context the Canal's role would be special since it can offer its services to the small sea–river-going vessels, not only because of the time economy it offers but also the safety it provides.

In order for the Corinth Canal to exploit the emerging intermodal market it is essential to proceed to short- and long-term plans that will allow for the re-engineering of its operation. As the market survey indicated, the key aspects of this reorganisation lie in the fields of pricing policy and the development of new services, while in the long term there is a need for new investment, in accordance with the dimensional characteristics and needs of the maritime market.

Finally, it should be mentioned that short sea shipping should not be considered as a microcosm of the traditional open sea transport. Short sea shipping has unique operational needs since it refers to the transport of goods for small distances and with small vessels. Short sea shipping could easily be considered to be a viable alternative to road transport and should be promoted accordingly as the most economic means in terms of private and social cost of transport. The Corinth Canal needs to revitalise its position through macro- and micro-economic objectives and actions in order to play an active role in the Mediterranean short sea shipping industry.

References

Commission Regulation (2000/C), 56/02 of 14 February 2000.
European Union COM (97) 243 final, 29 May 1997.

European Union COM (1999) 317 final.

Peeters, C., Verbeke, A., DeClerq, E. and Wijnolst, N. (1995), *Analysis of the Competitive Position of Short Sea Shipping: Development of Policy Measures – The Corridor Study*, Delft University Press, Delft.

Sambracos, E. (1988) 'Inland Navigation, a New European Proposal', *Hellas and National Transport*, 56.

Sambracos, E. (1999), 'The Role of Greece in the Inland Transport of Central and East Europe', 1st European Inland Waterway Navigation Conference, Hungary.

Sambracos, E. (2001) 'Introduction to Transport Economics', ed. Stamoulis, pp. 156–9 Athens (in Greek).

Sambracos, E. and Ramfou, I. (2000), 'The Role of the Corinth Canal for the Operation of the Greek Port System', 2nd National Conference on Port Works, NTUA, Athens (in Greek).

Tinsley, D. (1991) 'Short-sea Shipping, A Review of the North European Coastal Bulk Trades', London: Lloyd's of London Press Ltd.

University of Piraeus Research Centre (1999) 'Exploring the Possibilities for Broadening the Market of SA Corinth Canal', Research Program (in Greek).

PART III
IMPROVING MANAGEMENT AND QUALITY IN TRANSPORT IN SOUTHEASTERN EUROPE

Chapter 10

Organisation, Monitoring, and Efficiency of Athens' Urban Public Transport[1]

Alexandros Deloukas

Introduction

The future of urban public transport in Europe is a high agenda item, especially in view of the debate on efficient transport markets. The aim of the present chapter is the examination of the efficiency of the urban public transport in Athens, as well as of broad factors driving its performance. Two determining factors are investigated: firstly, the wider organisational arrangement of the urban public transport (UPT), and secondly, a performance management system producing strategic information on the efficiency with which input resources are transformed in transport services.

The regulatory structure of public transport in Europe is in an era of transition. As new regulatory regimes emerge, ongoing national and European Union-wide legislative initiatives arise. There are several reasons for such an evolution. Firstly, the closed markets in public transport (PT) have led to higher (than justified) subsidy levels for PT operators. Non-transparent cost accounting systems or cross-subsidisation of transport services are typical marks. Secondly, an arising refocus from 'public service' to 'service to the public'. The basic philosophy is simple. The fulfilment of transport needs should not be linked to the service provider (existing or new, public or private), but to the service provision itself. Third, experience with current reforms in certain European Union countries. Special attention is paid to the high transaction costs of full competition regimes.

Certain European Union-funded research studies investigated recently the pros and cons of different UPT market-structures,[2] as well as the tendering, contracting and performance monitoring in UPT with an emphasis on service quality.[3] Normative arguments for pure market regimes in UPT and more impartial analysis of transit authorities' practices provide useful insights about the current market and policy trends of the said subject. Corroborated research results are reflected in the recently proposed European Union Regulation (26

July 2000), concerning the award of public service contracts in passenger transport.

Other European Union transport research studies[4] examined external constraints of cost efficient operation and departures from efficient pricing of PT service provision, minimising social welfare losses. Benchmark comparisons of regulated firms, as well as mechanisms counteracting the propensity of subsidised utilities to realise higher (than minimum) costs are also instrumental in the UPT context.

Information technology in public transport operations can provide monitoring tools related to the service provision and also support strategic decision-making processes. The Enterprise Resource Planning (ERP) technology is able to integrate functional activities of an operator in a single software suite.[5] Measurement systems of Key Performance Indicators (KPIs) for operators may be developed, concurring with the balanced scorecard framework. The said framework is a leading strategic management concept linking a balanced set of performance results with driving processes of a firm.

Basic contributions of the chapter pertain to: (a) a benchmark comparison of PT operators in the Athens context; (b) novel development of IT-based practices monitoring PT operators; (c) a comparative description of closed market variants of PT in Greece.

The structure of the chapter is as follows: the following section contains a précis of the proposed European Union Regulation 7/2000 concerning public service contracts in transport. The chapter then presents comparative reviews of regulatory forms of UPT in Europe with case studies demonstration, and closed market variants of PT in Athens, Thessaloniki, and the rest of the country. Next, it addresses the pricing of PT service provision in general terms, as well as in the Athens context. Moreover, a yardstick comparison of similar PT operators in Athens enabling cost efficiency gains is conducted. The last part of the section discusses critical cost drivers of the metro operations. Following this, the development of an automated system to measure and predict the performance of the new metro lines 2&3 in Athens is elaborated. A Performance Management and Controlling (PMC) module, developed as an application embedded within an ERP system, is described.

The chapter concludes that, in view of the proposed European Union Regulation, there is a need for an independent regulatory authority of UPT in Athens. Good practice sharing within the Athens public transport system is suggested.

The Proposed European Union Regulation

The proposed Regulation applies, *inter alia*, to urban and interurban public transport. At least half of the public service contracts (by value) should be awarded following the provisions of the Regulation (amended as per 21 February 2002), within four years after its entering into force. The Regulation will be binding for all European Union member states. It lays down the conditions under which are granted to operators fulfilling public service requirements, operating subsidies and/or exclusive rights to operate (*concessions*). An introductory text of the proposal evaluates three types of PT regulatory regimes:

1 *full competition or deregulation*, in which no exclusive rights exist;
2 *controlled competition*, in which exclusive rights of fixed length are awarded following competitive tenders;
3 *closed markets*, in which operators do not face any competition and are protected by exclusive rights.

The ISOTOPE research study concludes that cost savings of about 15 per cent in unit operating costs are feasible in competitive tendering regimes over closed market structures, even without redundancies or wage reductions. On the other side, deregulated services are even cheaper, but tend to be substantially worse in terms of PT service quality. The controlled competition regime is endorsed by the proposed Regulation. A public service contract will be awarded giving exclusive rights (concessions). Contract length should be no more than five years, except if the operator provides a relevant volume of capital assets (vehicles, infrastructure). In the typical case, the contract should be put out for international, European Union-wide competitive tendering (least subsidy tender).

Direct award of contracts is foreseen for rail operators under very restrictive conditions. Efficient (benchmarked) bus services integrated to rail services may also be awarded directly (concerns integration of tariff, information, ticketing, timetables, use of interchanges). Note that, according to the German industry self-assessment, no German operator currently meets such requirements.

In all cases (competitive tendering, direct or quality-based award) the operator is accountable for its results, *inter alia*, it bears the commercial risk. When the subsidy level is not the result of competitive tendering, the net cost principle applies, whereby the cost calculation should be based on benchmark costs (not historical costs).

Comparative Review of Organisational Arrangements

Regulatory Forms of Public Transport in Europe

Full competition The deregulation of British bus services following the Transport Act 1985 is considered as an archetypal case study. Local bus services outside London have been fully privatised. Bus operators are free to supply services without exclusive rights to protect them. Each operator advertises and informs about its own services, being not interested in service or marketing integration with competitors. The barriers to market entry and exit are low. Many new (but small) entrants have been merged and the 1990s witnessed a strong concentration of the bus market.

According to the ISOTOPE study, full competition compared to other regimes is clearly superior in terms of productive efficiency (unit operating costs/veh.-km), but inferior in terms of consumption efficiency (unit costs/ pass.-km). The ridership declines over time due to poor service integration and instability of service supply (unstable interfaces, delays, bankruptcy risk). It is the case that public policy goals, such as traffic decongestion or environmental protection, are more difficult to achieve by this regime.

Due to incomplete information about the new entrants and the unknown element in operational characteristics, a regulatory authority is needed, even in a full competition regime. If some minimum safety and service quality standards are set, then the costs of monitoring and enforcing compliance, as well as handling disputes, coordinating operators, and handling intangible uncertainty, may become substantial. In some cases, full competition may require more regulation than a state monopoly (deregulation paradox), and the supervision costs may exceed the cost savings of the full competition regime.

Controlled competition With the Dutch Transportation Act from January 2000 onwards control over urban and regional transport was transferred from the central state to 35 regional provinces. The provinces formulated phased concessions which are tendered periodically. Competition among (public or private) PT providers prevails, when concessions are tendered. During the concession period, the selected operator has a regional monopoly. Apart from the concession definition (coverage area, length, performance metrics), the regional transit authority specifies in a so-called 'Schedule of Requirements' the required level of service, the subsidy level and the information to be delivered to the government. More control is exercised on decomposed input factors (e.g. labour or vehicle productivity), and less on the managerial autonomy of the operators and the complexity of bid evaluation. Greater focus on the control of generic output factors (e.g. minimum veh.-kms) increases the discretion of the operators and the complexity of bid evaluation. In the latter case *net cost* concessions prevail over simple *gross cost* contracts. Net costs are defined as the difference between gross costs and gross revenues. The Act allows the freedom of fare setting at the regional level. The concessionaire bears then the commercial risk. Competitive tendering is expected to reduce the government subsidy to a justified level, together with efficiency gains for the operators. The market, however, remains 'inactive' during the concession, so that a system of rewards and penalties is used to control the fulfilment of the prespecified performance metrics.

A *net cost* contract provides an internal incentive mechanism, in the sense that the operator benefits financially if it sells more transport service. It involves a higher (commercial) risk as opposed to gross cost contracts, which contain a lower (industrial) risk. Note that the proposed European Union Regulation 7/2000 opts for net cost contracts. In absence of external control mechanisms, operators being awarded a *gross cost* contract tend to reduce service quantity and quality. The (unfair) results are higher cost savings, therefore higher profits for the concessionaire.

Closed market The current UPT market in Greece resembles a closed market regime in which all operators are protected by exclusive rights and do not face competition from other potential entrants. The (public or private) monopolistic supply of services has led to inefficiencies, low service quality, and (surprisingly) poor level of service integration. The lack of competition pressure correlates with the risk of operators' complacency.

In the last three years three legislative initiatives have taken place to improve the service offered in Athens (Law 2669/98), Thessaloniki (Law 2898/01), and the rest of the country (Law 2963/01). The three Acts refer to local variants of the closed market regime.

In *Athens* the state delegates the UPT service for an unlimited period to a publicly-owned company, the latter being both sole owner of and transit authority for the subsidiary operating companies.

In *Thessaloniki* the state directly awards long-period exclusive rights to the incumbent private bus operator and subsidises it according to a performance contract till 2009. Its assets are passed afterwards to the state.

In the *rest of the country* direct awards of long-period concessions (till 2011) are made to the incumbent private bus operators. The (inter)urban bus operators do not receive operating subsidies, but capital subsidy for the modernisation of the bus fleet and the terminal infrastructure.

Athens Urban public transport in Athens is characterised through growing cumulative deficits of most operators (mainly due to closed market inefficiencies), rising externalities of car and taxi modes, and accelerated capital investments in UPT overall.

The last point needs further elaboration. In the last 30 years the Athens UPT system focused on the coverage of the transit-dependent population, mainly with bus services ('captive riders'). Since the operation of the two new metro lines in the year 2000, and in view of the high ongoing rail investments (metro extensions, light rail, suburban rail), a strategic change of focus embracing 'modal choosers' prevails. Choice riders wanting to avoid road congestion, have begun to appreciate the urban rail convenience, reliability and speed. New customer demands are slowly evolving on the UPT.

The Urban Transport Organisation OASA, being the Athens transit authority, became after the Law 2669/98 the sole shareholder of the existing operating companies ETHEL (thermal bus), ILPAP (trolley buses) and ISAP (metro line 1). The state is currently the sole shareholder of OASA. The three subsidiaries constitute the so-called Public Transport Provider Bodies (EFSEs). A fourth entity, AMEL (subsidiary of the Attico Metro company, owner of the metro system), operating the new metro lines 2 and 3 is also contained in the planning of OASA.

Along with gradually realising strategic planning responsibilities of OASA (revenue redistribution, introduction of bus priority measures, monitoring of transit service delivery, network modelling studies), most of its foreseen responsibilities are of tactical, even operational nature (bus routing and scheduling, installation of bus shelters and road markings, counts, passenger information). Other tactical and operational objectives, such as revenue collection and ticket control, transfer stations and ticketing system design, or operations control centre, are fulfilled by the operators themselves, Attico Metro, or the Athens 2004 Committee respectively. As yet unrealised strategic objectives of OASA refer mainly to the setting of service and quality standards or to the strategic marketing of the UPT image.

Thessaloniki Law 2898/01 established the Thessaloniki Urban Transport Council (SASTh) as the local transit authority undertaking certain planning and control responsibilities. SASTh's financial base (1 per cent of the bus operator's turnover) is under-proportionate compared to OASA, and also weak in absolute terms. The Law contains the bus service provision contract between the state and the private monopolist OASTh (Thessaloniki Urban Transport Organisation). The contract contains exact specifications of service output requirements in terms of veh.-kms and frequencies. The level of detail concerning allowable input resources and expenditures is striking. Beyond vehicle capacity obligations, minimum allowable productivity metrics per staff category are specified, as well as performance obligations concerning required productivity growth through time. Revenues are assumed to increase by 2 per cent per annum. Due to the subsidy formula, the maximum subsidy level depends almost exclusively to the administered fare level.

Advantages of the law are the (intended) financial discipline, and an innovative gain sharing plan for further cost reductions or revenue increases. However, OASTh compared to the EFSEs in Athens, has very limited managerial discretion in deciding how to allocate its input factors. It is paradoxical that it assumes commercial risk (due to the fixed subsidy level) with such a constrained managerial autonomy. Concerning service output, the Law emphasises service quantity and pays less attention to service quality.

Rest of the country Law 2963/01 covers the (inter)urban bus transport, leaving out Athens and Thessaloniki. The incumbent private bus operators build up so-called Common Bus Revenue Funds (KTEL) at the jurisdictional level of the 52 Greek prefectures. Each KTEL is essentially a joint venture society of bus owners, a large proportion of whom drive the buses themselves. There exist 93 regional KTELs, out of which 33 serve mainly urban areas and 60 interurban connections. The KTELs operate currently about 5,150 buses (of which 13 per cent belong to urban Bus Funds), engage 15,000 employees (urban funds: 20 per cent), carry 150m. riders per annum (urban: 85 per cent), and achieve an annual revenue of €460 m (urban: 75 per cent). Main features of KTELs' operation are their low unit cost of operation, independence from public subsidies, poor level of infrastructure and service comfort.

By means of financial incentives to the Bus Funds, the Law promotes the change of their legal form (incorporation) and old vehicle replacement (subsidised up to 20 per cent), as well as infrastructure investments, e.g., terminal stations (subsidised up to 50 per cent) within the remaining period until the 2004 Olympic Games. The KTEL incorporation is expected to improve access to private capital. The strongest incentive is the direct award of exclusive rights to existing (to-be-incorporated) KTELs to run their lines for a period of ten years at least. Competitive tenders are foreseen in cases of regional Bus Funds (with more than 12 buses) which do not wish to become incorporated.

In case of unproductive tenders, urban 'Public Service Requirement' lines (mostly revealing non-profitability) may be run by the local municipalities. The PT connectivity of over 5,000 rural villages recently merged to nearly 500 municipalities (by means of the administrative reform programme KAPODISTRIAS), if not covered by KTELs, is left to the municipalities themselves. However, no earmarked financial support is foreseen for this local 'service to the public' (as opposed to the subsidised metropolitan areas of Athens and Thessaloniki). It is expected that commercially viable (i.e., profitable) lines will continue to be operated by incorporated KTELs. Overall, the state has taken an important legal and financial initiative to push forward modernisation of the (inter)urban bus services. An extensive renewal of the regional bus fleet and infrastructure is anticipated.

Responsibilities of Transit Authorities in Europe

A survey of transit authorities in large European cities compiled by Lecler provides a basic source reference. The analysis is in terms of five dimensions: technical competence, revenue allocation, marketing, range of responsibilities and contractual relationship with the operators.

Technical competence Authorities may have technical competence on infrastructure investment and operation of all UPT modes (as in Stockholm or Lyon), may depend on operators for modelling studies (as in many German cities), or rely on experienced rail operators to operate services as they wish.

Revenue allocation PT organisations may collect fare revenues (as in Copenhagen or in German tariff associations) and refund operators according to allocation keys based on passenger surveys. By reliable surveys, this system increases the accountability of the operators concerning their market results. Alternatively, they may only organise the compensation schemes between operators for cash collected (as in Rhein – Ruhr or Athens). The latter way can be preferable, because it gives to the operators a stronger incentive for selling tickets.

Marketing Authorities may struggle to provide comprehensive information in deregulated environments (as in most British cities) or assume the strategic marketing of the PT image and delegate the product marketing to the operators (as in Zurich and Basle). German transit associations exercise effective forms of direct marketing, targeting new potential clients, who belong strictly to the segment of modal choosers. The case of Zurich is especially instructive. The canton and the local municipalities participate in the transit authority without involvement of the central (federal) state. The eight bigger participating operators are market-responsible for the sub-networks that they run, being allowed to subcontract certain routes. The zone-based tariff integration of all operators demonstrates also a best practice in this respect.

Range of responsibilities Transit authorities may be dedicated solely to public transport matters (as in the majority of the European cities) or build up a metropolitan authority responsible for (apart from public transport) land-use planning (Dublin, Helsinki, Rome), environmental control (Helsinki), metropolitan road network and congestion charging (London). The creation of metropolitan authorities with an extensive range of responsibilities for all aspects of urban mobility (especially traffic management and parking) and for land-use planning, is also a recommendation of the ISOTOPE research study. ISOTOPE considers metropolitan authorities as providing an integral framework for coherent policies that improve the urban quality of life. The participation of the communities directly affected is beneficial for the success of the metropolitan authorities.

Contractual relationship with the operators Apart from public service contracts for exclusive rights awarded through competitive tendering procedures (e.g., Frankfurt, Stockholm, most French cities outside Paris), transit authorities may close, in absence of competitive pressure, *management agreements* with operators. Such agreements make an attempt to reduce inefficiencies and excessive costs through a system of awards and penalties. Fairness requires a bonus or a penalty for the authorities too, in cases that they do or do not fulfil their commitments (e.g. provision of priority lanes to bus operators). A benefit of such agreements seems to be the definition of what is expected from the operators in terms of performance metrics. They make more sense when the authority defines output-based metrics and does not assume operational functions, e.g., prescribing input factors such as schedules and rosters. The European Union research study PRORATA[6] endorsed a quantified relationship for the rail transport: the more empowered and accountable the management for its commercial operations, the higher are the cost efficiency gains (or subsidy reductions) for the rail company. In that sense, it is more favourable for the authorities to focus on policy formulation and setting of standards than to interfere into the business of the operators.

Efficiency of Urban Public Transport in Athens

Pricing and Cost Efficiency – Optimal Pricing and Subsidy Level

Pricing It is part of the neoclassical school of thought in economics that social welfare is improved and resources are used in an optimal way when their product (e.g., provision of public transport services) is priced equal to its *marginal social cost*. Strict marginal social cost pricing prevails in the ideal world of perfect competition in related transport markets. Nevertheless, market failures are likely to happen. First, it is well known that the related real-world market of car use produces welfare losses. Car use is not priced at all at its marginal social cost, due to the high externalities that it produces. Secondly, it is also known that monopolistic operators (concentrating, say, more than 25 per cent of the public transport market, according to the proposed European Union Regulation 7/2000) in an unregulated environment by growing demand would try to overprice services instead of increasing their service frequencies, with resulting welfare losses. Regulation is needed to some extent therefore to correct or prevent such failures of (uncontrolled) markets.

Urban transport as a whole is currently underpriced in Athens, not reflecting its true costs. It is estimated that the hidden subsidies to car users due to externalities are much higher than the overt subsidies to UPT. More specifically, the underpricing of car use in Athens reduces the urban PT usage. In other words, inefficient pricing distorts modal competition within the urban transport market. The Greek government has been traditionally led by macro-economic constraints of inflation taming, income redistribution or fiscal pressure, when administering the fare level in the UPT market. This pricing policy contributed to the deficit growth of the operators and the degraded service quality provided. The latter's consequence was the almost exclusive attraction of captive riders revealing a very low fare elasticity of PT demand.

Setting market imperfections aside, there is a debate about the reference period of the marginal cost pricing of UPT service provision. In the short run (say, three months), the capacity of the operator is given and many cost categories are fixed. In the long run (say, three years), the capital stock is adjustable and most cost categories can vary. For publicly-owned, subsidised operators in a closed market environment, excess capacity is the rule – the short-run marginal costs are then lower than the long-run marginal costs (SRMC < LRMC). SRMC may be highly volatile over a year's period, or even more over a month's period. Because UPT pricing cannot fluctuate to match the SRMC, LRMC is a more stable basis for pricing operators' services.

Since the *rail* sector displays high fixed capital costs of infrastructure and rolling stock, in the typical case its average costs AC, first, decrease as output Q increases, (i.e., declining unit costs per additional veh..-km), second, are higher than its marginal costs MC. So, rail operators exhibit typically cost economies of traffic density, defined as

$$E_Q = \frac{MC_{\Delta Q}}{AC_Q} < 1 \text{ where total costs}$$

SRMC do not include capital costs which are fixed in the short run, so pricing P = SRMC will certainly lead the operator to financial deficits. LRMC–based pricing (P = LRMC), at least in the *rail* sector, does not recover the total costs TC, due to the existing economies of traffic density. In other words, the optimal or so-called *first-best pricing* based on marginal costs will lead, at least the *rail* operator, to financial losses. When economies of density exist, only long-run *average* cost pricing (P = LRAC) could recover the total costs of operation and achieve break-even (revenues = total expenditures). This is the rationale on which pricing and the 20-year Business Plan of Attico Metro are based upon.

Apart from reliability and service availability, quality of service in terms of comfort, cleanliness or security has its costs. The high service quality offered by the new lines 2 and 3 in Athens has also allowed the fare mark-up of the metro services compared to the other operators. Price discrimination based on service quality turns to be politically acceptable even in a closed market regime.

Alternatively (and as proposed by the CAPRI research study),[7] in view of the financial consequences of the first-best pricing, the government may provide, as a *second-best pricing* strategy, a subsidy to the operator (selected through a competitive tendering) in order to cover its deficit, and require that it prices at the marginal cost, i.e., minimising the loss of efficiency.

Subsidy level In exchange for government subsidies, PT operators fulfil public service and tariff requirements. Beyond the self-evident compliance to safety standards, the PT service requirements are interrelated and refer to *service quality* (e.g., reliability, lack of cancellations, cleanliness, security), and *produced output* (e.g., frequency, veh.-kms, seat-kms). Note that seat-kms do take account the vehicle capacity involved. In Switzerland and in many German tariff associations, operating subsidies are also related to the *consumed output* (e.g. passengers carried, travel-cards sold) in order to stimulate ridership growth, modal split improvement and reduction of car externalities (second-best pricing strategy).

In Athens, the Law 2669/98 prescribes that OASA receives from the state a lump-sum amount (overall subsidy), based on the budgeted overall *net costs* of PT operations (= difference of overall PT operating costs and revenues). However, the lump-sum amount is independent of the performance of EFSEs or of OASA as the parent company. The budgeted 'next year' subsidy is estimated considering the fare policy, the targeted change of the PT operating costs (a percentage increase should be less than the inflation rate), and the anticipated change of the PT operating revenues. In the common case that the subsidy is not fully covered by the state budget, the state provides a guarantee for new loans raised by OASA. The state has written-off penalties on old debts of the EFSEs for overdue loans, and the equity share of OASA in the EFSEs has been increased to the same amount. OASA compensates for the time being (through the overall subsidy and the raised loans) the net costs of each EFSE. Law 2669/98 foresees however, that OASA will compensate each EFSE according to some, as yet not specified criteria (including Business Plans, management agreements, output and quality measures).

Better net results can be achieved by an operator by lower costs (efficiency of operations) and/or higher revenues, i.e. higher ridership by administered price (effectiveness of operations). Law 2669/98 considers each EFSE operator essentially as a cost centre, not accountable for the effectiveness of its operation. However, in the service industry, such as the passenger transportation, it is difficult to separate marketing and promotion from the production. The service production and consumption process cannot be unbundled and its component functions cannot be handled separately without seriously altering its effectiveness.

The Law, being captured in the 'cost centre' perspective, does not provide any mechanism to allocate joint (e.g. travel-card) revenues among operators. A clear separation of transparent expenditure and revenue accounts *per operator* is required, otherwise a cross-subsidisation of mismanaged entities through more efficient operators is not precluded.

The level of the overall subsidy of the UPT system in Athens has been found to be excessive, i.e., higher than justified.[8] A real problem is that the minimum (i.e., necessary) costs of each operator, given its output level, are unknown. Even the actually incurred (i.e., necessary *and* unnecessary) costs are not transparent. The estimation of avoidable costs of operation and of a justified subsidy level is, therefore, disabled. Note that every operator has a different cost function. What is needed is a mechanism to counteract the propensity of each operator to overstate minimum costs and/or to waste input factors by letting higher costs to incur (since in each case a higher subsidy results). The latter technical (so-called X-) inefficiency concurs with the empirical evidence that a firm working under a *closed market* structure, given its output level, fails to minimise its total costs of production.[9] For instance, it may pay for a given output level higher input prices and/or more input quantities than are necessary.

A simple mechanism counteracting the above mentioned behaviour is an adaptation of the Vogelsang–Finsinger.[10] The regulatory authority does not need to know the cost function of each operator, but cost accounting data only. The operator is able to have any average income (= revenue + subsidy) in the 'next year', as long as this average income multiplied by the output (e.g., veh.-kms) in the 'current year', does not exceed the operator's costs in the 'current year'. Successive application of this constraint over several years should result in equilibrium, whereby the average income equates the average costs. The above mechanism promotes higher financial discipline and performance improvement, both for mismanaged entities as well as for more efficient operators.

The proposed European Union Regulation 7/2000 advocates the *controlled competition* regime for the provision of PT services. Under this regulatory form, competitive tendering of PT services prevails. Competitive tendering is an efficient way of ensuring that operator's subsidy is *not* excessive. The contractual compensation of the concessionaire equals his net costs. A *net cost* contract increases the accountability of the operator and provides an incentive for ridership growth.

Benchmark Comparison of Operators

The subsidy scheme of Law 2669/98 for covering operating losses does not address the problem of an efficient cost reduction by the operators. The latter have no worthwhile incentive to minimise or simply to reduce their costs. The state (or the regulatory authority) does not know what the appropriate cost level should be. This is a classical case of asymmetrical information, where the operator knows more about its costs and the government possesses insufficient information. What is needed is a *benchmark* other than the operator's past performance, against which to assess his potential for cost reduction.

One simple method is a cost comparison across similar operators. All four PT operators in the Athens area share the same regulatory environment and regional cost-of-labour index. Metro and bus operators are dissimilar across technology, infrastructure, and exclusive rights-of-way, so they may be considered as not comparable. The pairs of bus operators 'BO' (thermal vs. trolley buses) and metro operators 'MO' (line 1 versus lines 2 and 3) may be considered as internally comparable, despite any differences in the coverage area, service level and vehicle types. Such differences are not really exogenous. They can be altered through upgrading investments (e.g., dual mode trolley buses) and policy measures (e.g., bus lanes), which can potentially lead to cost reductions.[11] However, certain genuine differences do exist in Athens within each comparison pair and pertain to: (a) catenary-related costs between both bus operators; (b) underground system-related costs, mainly station power and maintenance costs, between both metro operators. In both cases a correction of the should-be unit cost level is allowable.

To repeat, the state (or the regulatory authority) does not need to know the exact cost function of any operator: what is needed are cost accounting data only. The relative performance within each pair can be used as an instrument enabling competition and cost control.

Two different average unit cost levels (operating cost per vehicle-km/peak vehicle) and a staff productivity indicator (employees per peak vehicle, as a critical cost driver) are presented in the tables below. Note that the peak period capacity requirements condition the necessary fleet size. Operating cost figures do not include capital costs.

The three performance indicators for the bus operators (named, for the sake of the argument exposed herein, 'BO1' and 'BO2' respectively) are based on the year 2000 figures of the OASA/EFSEs 2001 revised budget report (September 2001).

Bus performance indicators	BO1	BO2
Operating cost per vehicle-km (€)	4.88	226.19
Operating cost per peak vehicle ('000€)	7.62	2.05
Staff per peak vehicle	132.97	4.02

The same performance indicators for the rail part of the metro operators (named 'MO1' and 'MO2' respectively) are based on the year 2000 figures of the same report.

Metro performance indicators	MO1	MO2
Operating cost per vehicle-km (€)	3.85	378.51
Operating cost per peak vehicle ('000€)	7.37	2.91
Staff per peak vehicle	315.86	5.23

The benchmark comparison demonstrates that 'BO2' and 'MO2' are the (more) cost efficient operators within their group. Note that the comparison is not intended to grade operators, but to enable the sharing of better practices.

Within each comparison pair, the underperformer can be urged to achieve the unit cost level of the more efficient. Due to the significant difference of the estimated unit cost levels, such a performance improvement would imply substantial subsidy reductions and social welfare gains.

Cost Structure of Line 2 and 3 Operations

The annualised capital costs of the UPT operators are proportional to the summed depreciation rate and interest rate of their capital stock. Capital costs are of course much higher for rail than bus operators. Bus operators (as opposed to rail) do not pay maintenance and capital costs for the physical network that they use.

In the short run (say, a three month period), concerning metro operations, the capital costs are considered as fixed. Other input factor costs, not adjustable to equilibrate a changing output (e.g., veh.-kms), are also regarded as fixed. Considering part-time and overtime salary costs, traction power costs, rolling stock maintenance materials consumed, and outsourced costs (cleaning, security, maintenance support contracts), as costs linked to the level of metro service, then the *variable costs* of metro operations amount to 34 per cent of the total operating costs (excluding capital costs). Note that certain variable costs (e.g., ticket collection) are *substitutes* for fixed costs of technology-intensive capital stock (e.g., automatic fare collection equipment).

The *outsourcing ratio* itself amounts to about 23 per cent of the total operating costs (last-12-month moving average by June 2001). In terms of cost flexibility and avoidability, the ratio indicates a high performance of the metro operation. Outsourced activities contribute to the metro service quality, and are critical cost drivers of the metro operations. Some elaboration seems advisable. *Outsourcing* of an input activity to a third party is a lean technique to convert fixed costs, and particularly fixed labour costs, into variable costs. It enables a higher cost transparency, as well as a better performance control. Performance-based contracts based on competitive tendering are the rule for the line 2 and 3 operations. The cost advantage of outsourcing reduces however, with an increasing system size Q of the metro operations (e.g., number of peak vehicles or stations), as demonstrated in Figure 10.1 for an outsourced activity (e.g., security).

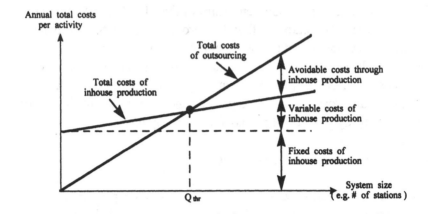

Figure 10.1   Outsourcing versus insourcing decision

Q_{thr} is the minimum threshold for which in-house production is more economical than outsourcing. In-house production or insourcing would mean essentially a diversification of the operator to non-core activities. A third, intermediate option would be an 'outsourcing' of an activity to a dedicated subsidiary.

Performance Monitoring for Metro Operations

Assessing the performance of a public transport operator with its multiple activities has become an important concern in the last decade. Expectations of the stake-holders (passengers, owners, employees, community) for capital-intensive metro systems are raised to a high level. Within this context, an automated system to measure the performance of line 2 and 3 operations in Athens has been developed.

Starting from strategic initiatives assembled within the 20-year Attico Metro Business Plan, relevant business strategies are transformed into 21 partial objectives of critical importance, belonging to four perspectives. The framework provides a balance between financial outcomes of the operations (e.g., cost recovery ratio) and the drivers affecting the future performance of those outcomes. A cause-and-effect relationship has been hypothesised about the four perspectives. *Employee satisfaction* has been assessed as a very basic resource of *efficient business processes* (e.g., productivity growth). The latter processes determine *quality of service* satisfying the passengers (e.g., service reliability). The level of service along with the efficiency of business processes have, finally, a strong impact on the future *financial effects* of metro operation.

The measurement system of the metro performance has been selected to match the structured methodology of the balanced scorecard framework[12]. The measures, termed Key Performance Indicators (KPIs), interrelate input resources and factor costs (labour, energy, materials), service production (e.g. veh.-kms, quality level), as well as service consumption variables (e.g. ridership, revenue), as shown in Figure 10.2.

Input resources/OPEX/factor costs
- Labour (staffing)
- Capital (e.g. #rolling stock vehicles, materials)
- Energy (traction power)

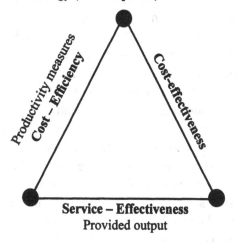

Service – Effectiveness
Provided output

Service production
- Vehicle – kms
- Seat – kms
- Service availability
- Service reliability

Service consumption
- Passengers
- Passenger – kms
- Farebox revenue
- Safety (accidents)

Figure 10.2   Conceptual inter-relationship of market variables

The *financial* KPIs (6) pertain to cost efficiency (operating costs/veh.-km), cost effectiveness (operating costs/passenger), outsourcing ratio (outsourced expenses/operating costs), revenue ratios (fare-box revenue/ridership, ancillary revenues/operating revenue), and cost recovery ratio (operating revenue/operating costs). The *customer-oriented* KPIs (7) refer to train service availability (mileage operated as per cent of scheduled mileage, percentage of peak train cancellations), train service reliability (actual train runs with delay <2 min as per cent of scheduled runs, failures causing delays/10,000 veh.-kms), in-vehicle peak crowding level (standees/sq. m.), station service quality (escalators service time operated as a percentage of planned service time), and safety (passenger injuries/10,000 veh.-kms). The *business efficiency* KPIs (6) pertain to labour efficiency (employees/10,000 veh.-kms, employees/route-km, train-kms/driver, rolling stock maintenance employees/vehicle), peak vehicle utilisation (peak vehicles/active vehicles), and vehicle efficiency (veh.-kms/ vehicle). Finally, the *employee satisfaction* KPIs[13] refer to the turnover rate and the non-attendance rate.

The AM Business Plan established a valid comparison with best practices of competitive European metro systems (Barcelona, Vienna). The derived target benchmarks represent a demand, such as a significant reduction in input factors utilised or a substantial increase in quality achieved. The KPI system assigns a quantitative measure to each strategic objective and tracks its value against the target benchmark. Line 2 and 3 operation exceed the target benchmarks set in many KPI dimensions. One of the reasons of the good performance is the high degree of outsourcing non-core activities.[14]

The role of information technology is essential for operators in order to: a) monitor operational processes related to service provision; b) support strategic decision making processes.[15] The ERP system integrates activities of finance, human resources, maintenance, warehouse and procurement based on a single database. A sequential implementation strategy has been adopted. At a first stage, business processes have been engineered along with the ERP customisation for the warehouse and maintenance activities. Especially the latter ERP module is considered as a novelty for European metros. In a subsequent phase, the rest of the ERP modules have been developed. Since metro divisions can share information and communicate with each other, the introduction of ERP technology is expected to lead to improvements in the future metro performance. For instance, warehouse shortages are revealed in real time, so inventories may be reduced. Financial reporting becomes more transparent and detailed, so that accountability increases. The necessary standardisation of processes enables a further productivity growth.

A Performance Management and Controlling (PMC) module has been developed as an application embedded within the ERP. The PMC module enables the automated monitoring of the overall performance of metro operations. It supports strategic management rather than operational processes. Basic functionalities of the PMC module are:

1 calculation of KPIs;
2 comparison with target benchmarks;
3 monitoring of performance variability over time;
4 prediction of performance through 'what-if' scenarios calculation.

The sourcing scheme for calculating KPIs is demonstrated in Figure 10.3.

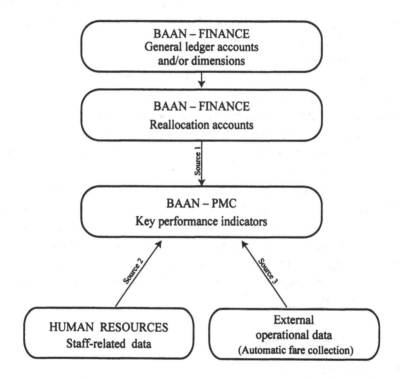

Figure 10.3 Sourcing scheme for calculating KPIs in the PMC module

Financial data of general ledger accounts and dimensions are consolidated through the reallocation submodule, staff-related data are drawn from the human resources module, and external operational data are entered through 'cold links' from the automatic train supervision system, train scheduling software, or manually. Note that the detailed financial accounts are regrouped in a manner that transcends the simple book-keeping function in order to support a strategic decision-making. Enterprise models define the interrelationships between the KPI components. Cause-and-effect or fishbone charts describe, within the PMC module, the multiple components ('causes') of an activity measure ('potential problem' or 'effect').

Findings and Conclusions

Three regulatory regimes of the UPT, associated with different political economic schools, have been identified. In the *full competition* regime, following the tradition of Friedrich Hayek, the spontaneous market forces should be left to prevail in the UPT sector without caring about service instability, operators' insolvencies or equity matters.

The currently prevailing *closed market* structure follows a corporatistic model. It enables a visual alert by underperformance, so that an exceptions analysis and corrective actions are possible without delay (early warning).

One value-adding functionality of the PMC module is that it not only monitors past performance, but also predicts future performance through 'what-if' scenario calculations.

Scenarios may refer to altered train schedules, work rules or added extensions, i.e., changed outputs such as veh.-kms, train-hours, peak hour vehicles or route length. The base algorithm is a calibrated unit-cost allocation model of metro operations. Cost predictions are a function of input prices an fluctuating outputs. The costing scenarios are run by the reallocation sub-module. Thus PMC capability goes beyond business reporting and enables business planning.

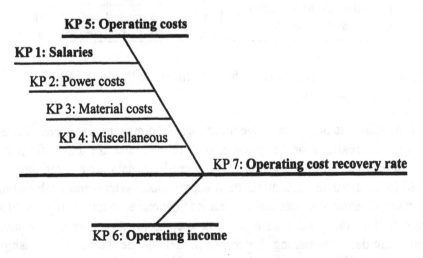

Figure 10.4   Linking of KPIs with the fishbone diagrams (Ishikawa)

It is characterised by cost inefficiencies and, in the Athens variant, by growing cumulative deficits and low service quality. The *controlled competition* regime concurs with the neoclassical model and is advocated by the proposed European Union Regulation. It acknowledges the need for competition in the UPT market, the price distortion because of the car externalities, as well as equity matters (e.g., carless households, mobility disabled, and non-commercial connection to low-density locations, all requiring a sufficient public transport service).

Most European countries are prepared or are preparing themselves actively for the proposed European Union regulation.[16] Denmark, Sweden, Finland, Holland, and many French and German cities already implement the principles of the proposal, and Italy is also preparing in this respect. Germany recently witnessed a merge of large transit agencies (Cologne and Bonn) to help withstand international competition better, when the European Union-wide tendering of services will be enacted.

One conclusion of the present chapter refers to the current organisational arrangement of the UPT in Athens. The existing legal and institutional framework of the Athens UPT is inadequate to meet the requirements of the proposed Regulation. Law 2669/98 does not distinguish institutionally between regulation and provision of urban public transport services, in so far as OASA becomes owner of the EFSEs. It is consequential that the Law does not make any provision for competitive tendering of UPT services. The arm's length relationship between the transit authority and its EFSE subsidiaries would reveal conflicts of interest in a *controlled competition* regime.

Benchmark competition is a way of improving the performance of the operators. The benchmark comparison carried out demonstrates that there is a space for better practices among operators in Athens. Flexible cost structure and high degree of outsourcing, as well as automated performance monitoring of strategic processes, are demonstrated in the case of metro lines 2 and 3. These may be rated, within an international context, as good practices.

In conclusion, there is a need for a regulatory reform of the Athens UPT, and in particular for the setting up of an independent regulatory authority. This would supervise competitive tenders of UPT services, as well as the adhesion to contract stipulations. Note that the new LRT under construction in Athens will operate on a private basis before the 2004 Olympic Games. The related public service contract will be the result of a competitive tender. The current transit authority in Athens, as long as it remains owner of the EFSEs, can guarantee neither impartiality nor neutrality towards new potential UPT providers. The growth of the cumulative deficits of the EFSEs signals also the need for sharing of good practices and for a more competitive market structure.

Notes

1   The views expressed in this chapter do not represent official views of Attico Metro AE, the author being solely responsible for views expressed herein.
2   ISOTOPE Transport Research Program, Imposed Structure and Organisation for Urban Transport Operations of Passengers in Europe.
3   QUATTRO Transport Research Program, Quality Approach in Tendering Urban Public Transport Operations.
4   Such as PRORATA and CAPRI.
5   See Schwartz, 2000.
6   PRORATA Transport Research Program, Profitability of Rail Transport and Adaptability of Railways.
7   CAPRI Transport Research Program, Concerted Action on Transport Pricing Research Integration.
8   See Deloukas and Karlaftis, 2000.
9   See Liebenstein, 1966.
10 This mechanism described in Train, 1992.
11 There is empirical evidence that bus operators exhibit U-shaped fleet economies of scale and, more precise, diseconomies of scale above an optimum fleet size of about 300 to 500 buses (see Berechman, 1993). The transferability of this evidence in the Athens context needs a closer examination, and if this proves to be the case, then the trolley bus operator is about the optimum fleet size. Under *ceteris paribus* conditions, a split of the thermal buses into three to four subregional operators would imply cost advantages and enable yardstick competition among them.
12 Olve et al., 1999.
13 Invented by Kaoru Ishikawa.
14 A detailed analysis is performed in Deloukas and Apostolopoulou, 2002.
15 A proprietary Enterprise Resource Planning System went live in AM on 2001, covering important business processes. The ERP standard software BAAN IV, based on the ORACLE 8 database platform, is licensed for more than 150 end-users (Wenzel and Post, 1998).
16 See Lecler, 2001.

References

Berechman, J. (1993), *Public Transit Economics and Deregulation Policy*, Elsevier, Amsterdam.

Cantos Sanchez, P. (2001), 'Vertical Relationships for the European Rail Industry', *Transport Policy*, 8, pp. 77–83.

CAPRI Transport Research Program (2001), 'Concerted Action on Transport Pricing Research Integration', Final Report, January, Contr. No. ST–97–CA–2064.

Deloukas, A. and Apostolopoulou, E. (2002), 'An Automated Performance Measurement System for Metro Operations', paper presented to the European Transport Conference.

Deloukas, A. and Karlaftis, M. (2000), 'The Social Costs of Auto and Transit Travel: A Total Cost Approach for the Athens, Greece, Metropolitan Region', paper presented to the 79th Annual Meeting Transportation Research Board, Washington DC.

European Commission (2000), 'Proposal for a European Regulation on Action by Member States Concerning Public Service Requirements and the Award of Public Service Contracts in Passenger Transport by Rail, Road and Inland Waterway', July.

ISOTOPE Transport Research Program, Imposed Structure and Organisation for Urban Transport Operations of Passengers in Europe (1997), *4th Framework Program*, Report 51, Office for Official Publications of the European Community, Luxembourg.

Jochem, M. (1998), *Einführung Integrierter Standardsoftware: ein Ganzheitlicher Ansatz*, Lang, Frankfurt.

Lecler, S. (2001), 'What Role for Public Transport Authorities in the European Metropolitan Areas?', paper presented to the European Transport Conference, Cambridge.

Liebenstein, H. (1966), 'Allocative Efficiency vs. X-Efficiency', *American Economic Review*, 56, pp. 392–415.

Monami, E. (2000), 'European Passenger Rail Reforms: A Comparative Assessment of the Emerging Models', *Transport Reviews*, 20, pp. 91–112.

Mouwen, A.M.T. (2001), 'Tendering and Contracting Public Transport in the Netherlands', paper presented to the European Transport Conference, Cambridge.

Olve, N.-G., Roy, J. and Wetter, M. (1999), *Performance Drivers: A Practical Guide to Using the Balanced Scorecard*, Wiley, Chichester.

PRORATA Transport Research Program, Profitability of Rail Transport and Adaptability of Railways (1999), *Final Summary Report*, July.

QUATTRO Transport Research Program, Quality Approach in Tendering Urban Public Transport Operations (1999), *Final Summary Report*, June.

Schwartz, M. (2000), *ERP-Standardsoftware und Organisatorischer Wandel*, Gabler, Wiesbaden.

Shleifer, A. (1985), 'A Theory of Yardstick Competition', *Rand Journal of Economics*, 46, pp. 319–27.

Simpson, B.J. (1996), 'Deregulation and Privatisation: the British Local Bus Industry Following the Transport Act 1985', *Transport Reviews*, 3, pp. 213–29.

Train, K.E. (1992), *Optimal Regulation: The Economic Theory of Natural Monopoly*, MIT Press, Cambridge.

Van de Velde, D.M. (1999), 'Organisational Forms and Entrepreneurship in Public Transport', *Transport Policy*, 6, pp. 147–57.

Wenzel, P. and Post, H. (1998), *Business Computing mit BAAN*, Vieweg, Braunschweig.

Chapter 11

The Use of Smart Card Technology for Efficient Transport: The Case Study of Thessaloniki

Georgia Aifadopoulou and Vasilis Mizaras

Introduction – Objectives

Smart card technology has been available for decades; however, it is only fairly recently that smart card-based applications have gained momentum and their adoption has become feasible. The current smart card status has been established due to a series of factors such as technical advancements, marketing maturation, hardware price decline, and, most importantly, concentrated research efforts. The transport sector plays a central role in the smart card micro-world, and for this reason a significant amount of the relevant research has been oriented towards the implementation of smart card applications in transport operations.

Within this context, the current chapter has two objectives:

- to provide a general, introductory information on the smart card concept and its use in transport operations. Furthermore, to catalogue identified benefits of smart card use;
- to present a resumé of research work on this area, by examining the case study of the Thessaloniki test site.

Overview of Smart Card Technology

Basic Facts

Smart card technology has a history going back at least 25 years. The French have been identified by most as being the concept generators, since the base technology patents belong to the French companies Bull and Innovatron.

The innovative element of the smart card was the concept of a single chip microcomputer; that is, processor plus memory all on one chip. Thus the smart card may be considered in the same light as a PC, but all on one chip and with externally supplied power, clock and input/output.

The original smart card chips were difficult and expensive to make. Over the years there have been many improvements in both the technical and price performance of smart card chips with the result that smart cards have become a common solution for many applications.

Smart Card Types

Overall, the following smart card types currently exist:

- memory cards, which contain just memory without (or with limited) processing power;
- contact smart cards, which must be inserted into a reader in order to communicate with the outside world through a metal contact plate on the card;
- contactless smart cards, which use radio-frequency radiation both to power the card and to transport data to and from the card. With this type of card it is not necessary to insert the card into a slot, just to pass it near to the reader;
- combi cards, also known as dual interface, which contain both a contactless and contact interface;
- program loadable cards, which are more advanced smart cards that are able to dynamically load programs into the card memory and execute them similarly to a standard PC.

Smart Card Applications

The prime mover of smart cards has traditionally been the telecommunications industry, which very early realised the potential of and adopted the new technology. The public has been familiarised with smart (or better chip) cards through the introduction of chip telephone cards used for public phones and chip – SIM – cards used for GSM mobile phones.

The finance industry, i.e. banks, has also been a significant target group for smart card manufacturers. The basic features of smart cards, such as portability and advanced security, suit the particular requirements of the financial industry well. The smart card technology is expected to take over the role of credit/debit cards from magstripe cards, but also to be the 'host' of new bank products, such as e-purses.

Smart cards are also used in a wide range of other sectors' applications. To mention just a few: in government as a secure means of identification (and probably replacing ID cards and passports in the future), in retail as loyalty cards, and in healthcare as a secure storage of patients' data.

However, the most promising smart card adopter sector seems to be transport, since its operations address a wide audience, which means a massive volume of cards and massive volume of daily transactions.

Perhaps the most advantageous feature of a smart card is the capability of carrying multi-applications. In other words, a smart card is suitable for hosting more than a single application at any given time. This has given a new sense to the terms integrated and/or interoperable ICT (information communication technologies): integrated systems very often base their operation on the use of smart cards. Transport applications within the scope of a multi-application smart card are usually referred to as the 'killer application', due to their magnitude in terms of targeted audience and the potential volume of transactions.

The Use of Smart Cards in Transport

Scope of Smart Card Usage for Transport Services

Smart cards are used for a wide range of transport purposes (and in various transport modes). Smart card transport applications include:

- electronic payment/ticketing in public transport;
- electronic payment in interurban road tolling;
- electronic payment/authorisation in urban road pricing schemes;
- electronic payment/authorisation in parking;
- vehicle/driver access control;
- ticketing, check-in and baggage handling for air transport;
- tachograph logging and management;
- customs and transit documentation.

Although, all of the above-mentioned have a significant share in the smart card market – for example, the German air-travel company Lufthansa has almost 800,000 smart cards in circulation – the most significant is public transport, although within a few years, road tolling may become equally significant.

Also, a parking application may well accompany road tolling and/or public transport in the framework of a multi-application smart card.

The scope of the current chapter is limited to the above-mentioned identified as most significant applications, namely public transport, tolling and, to a lesser extent, parking.

Types of Smart Cards Used in Transport

Although, the contact type is the most common sort of smart card generally in use, public transport (and to some extent tolling) requires faster and more user-friendly transactions and for these reasons contactless cards are preferable.

Combi-cards have emerged as a result of the required cooperation between financial applications (strong requirement for contact cards) and transport applications (strong requirement for contactless cards).

Benefits from the Use of Smart Cards in Transport

The introduction of smart card schemes by transport operators offers a wide range of financial and operational benefits leading to improved transport efficiency. The most important benefits are:

- introduction of more complex tariff policies, since the intelligence of the smart card can deal with the desired complexity of charging;
- faster transport operation; a single smart card transaction has a duration of less than 1 second;
- user-friendly transaction.

The above-mentioned potentially lead to:

- increased patronage;
- reduction of fraud; the smart card cannot easily be duplicated or tampered with and ticket validation is much more accurate;
- efficiency in ticketing management;
- improved knowledge of passenger traffic for planning purposes; all passenger traffic is electronically logged and stored.

In the framework of a multi-application transport smart card scheme, there are some extra benefits worth mentioning:

- facilitation of park-'n-ride facilities: the same smart card can be used for payment of both parking and public transport, with reduced fares for transit drivers changing to public transport;
- facilitation of congestion pricing schemes: the same smart card can be used for payment of both urban tolls and public transport, with reduced fares for transit drivers changing to public transport

The multi-application smart card has a higher level of market penetration leading in the end to its utilisation by non-captive public transport users as well.

The main conclusion is that smart cards, as well as performing the function of an ordinary ticket, could be used as a powerful demand management tool, which could assist local and regional authorities to formulate their transport policies, such as modal shift and environmental considerations. This assumption is proved through the conduction of surveys and measurements in the city of Thessaloniki.

Market Situation

The use of smart cards in transport is not yet a common practice among transport operators around the world. In 1999 only 885,000 public transport smart cards were in use in Europe (DISTINCT, 1999a). However, it is envisaged that transport cards will follow the tremendous upward trend of general smart card use: smart cards in circulation will quintuple from 1 billion in 1999 to 5 billion by 2005 (Frost and Sullivan, 1999). Accordingly, public transport cards existing in Europe are estimated to be 45 million by 2003 (DISTINCT, 1999a).

As far as public transport is concerned, the technology has already been implemented and is in use or expected to be operational in the next couple of years in several cities around the world, such as London, Paris, San Francisco, Washington DC, Seoul and Hong Kong. The large size of the city is not a prerequisite for the successful roll-out of a smart card scheme. For example, currently smart card public transport ticketing is operating in almost all Finnish cities except its capital city Helsinki (which nevertheless will soon follow).

Tolling operators also use million of smart cards, but at the moment only for the purposes of stop-'n-go payment, and not as a part of a non-stop fully automated system. For example, the Greek toll operator Attiki Odos introduced pre-paid smart cards for this purpose in 2003.

The Case Study of the Thessaloniki Site

Brief History of Thessaloniki's Involvement in Telematics Projects

The Greek city of Thessaloniki has been very active in telematics applications, both in national and European programmes. The city has an advanced telematics infrastructure in the sectors of transport, urban, tourism and healthcare. Some of these tools are implemented in the framework of the projects: CONCERT (Cooperation for Novel City Electronic Regulating Tools), ADEPT I and II (Automatic Debiting and Electronic Payment for Transport), QUARTET PLUS (Validation of a European urban and regional integrated road transport environment based on open system architecture) and DISTINCT (Deployment and Integration of Smart Card and Information Networks for Cross-sector Telematics). The main objective of these projects (sponsored by the European Commission Directorate General XIII) is the use of tools to make transport operations – with the emphasis on public modes – more efficient.

The current chapter will examine the results of the projects ADEPT, ADEPTII and DISTINCT, which have conducted research in the area of smart cards in Transport.

Presentation of the ADEPT, ADEPT II and DISTINCT Projects' Objectives

ADEPT (Automatic Debiting and Electronic Payment for Transport – 1992– 94) is one of the leading projects in the area of smart card and transponder-based systems for paying transport services. The project is sponsored by the DRIVE EC programme. Within the framework of ADEPT, the Thessaloniki site demonstrated a smart card-based, multi-lane tolling application at the Malgara toll station.

The ADEPT II project (1994–96) was the successor of ADEPT project and was sponsored by the Telematics Application Programme (TAP) of DG XIII in the transport sector. The ADEPT II project involves three main demonstration sites: Thessaloniki (Greece), Gothenburg (Sweden) and Helsinki (Finland) which demonstrate smart card and transponder-based systems to pay for several transport services, such as road tolling, booking and paying for parking, public transport payment, and others. The Thessaloniki site in particular introduced a multi-application, hybrid smart card that could be used for payment for tolling, public transport and parking.

Within the DISTINCT project (1998–2000), the Thessaloniki site (OMPEPT, Aristotle University of Thessaloniki, Region of Central Macedonia,

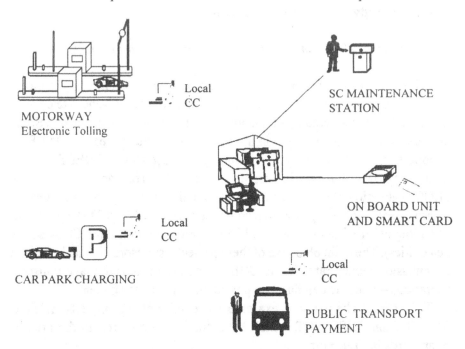

Figure 11.1   Overall design of Thessaloniki's integrated payment system within the ADEPT II project

Source: ADEPT II.

National Fund for Highways, INTRACOM SA, BIOTRAST SA and TRUTh SA Consulting) aims at the integration of the available applications in different telematics sectors both in the city scale and across the other DISTINCT sites (Lapland – Espoo – Vantaa in Finland, Newcastle upon Tyne in the United Kingdom, Turin in Italy and Zeeland in the Netherlands).

Research Context

The Thessaloniki site keeps a continuous record of these projects' results: it analyses and tests the effect of integrating existing applications and deploying multi-function smart cards on the level of service and the general improvements in the traffic congestion and, hence, air quality. The smart card initially developed for only tolling applications (ADEPT project) is being advanced to offer transport-related services such as public transport, parking and tolling payments (ADEPT II project), and is then being extended to a city card, which also includes health records and access to citizen and tourist information (DISTINCT project). The idea behind the integration work is to encourage the use of smart cards in transport and increase the number of users of public transport. Furthermore, existing traffic information networks (such as the VMS network) have been integrated with other networks and, thus, are available through other media such as info-kiosks and cellular phones. As a result, traffic information is disseminated to a wider range of potential drivers enhancing the impact on their travel behaviour.

Assessment Hypothesis

In order to assess the potential transport and environmental impacts from the introduction of integrated applications, and for the wide range of users group involved, the site has performed extensive surveys and interviews in the framework of the ADEPT and DISTINCT as well as other projects' evaluation activities. Some of these data are presented in this chapter, and they are correlated with the generic identified smart card benefits also mentioned in this chapter.

The assessment hypothesis have been set as following:

- improved transport efficiency – smart card introduction will have positive impacts on the city's transport operations;
- increased market penetration – end users are willing to acquire and use smart cards, but the multi-application feature of the smart card will further encourage smart card penetration;
- financial viability – smart card implementation investment will have a high rate of return.

Furthermore, the evaluation work identifies the drawbacks related to the introduction of the multi-application smart card.

Improved Transport Efficiency

The implementation of smart cards as a means for electronic payment of public transport consists a vital supporting measure for traffic demand management and modal shift, due to: a) friendliness of technology to the end user; b) migration from the single-trip ticket to the multi-trip ticket concept; and c) the 'marketing' opportunities emerging for the public transport operators. According to the user surveys conducted in May 1998 in Thessaloniki, it was estimated that the increase of public transport usage due to the smart card introduction for public transport payment alone would be at the range of 3 per cent. As a logical step, the Thessaloniki smart card also facilitates electronic payment for tolling and parking, thus constituting a generic transport payment card. As a result the smart card is used as a lever to disseminate the 'park and ride' concept to the general population and, moreover, is penetrating the drivers' user group as well rather than solely the public transport users group. The same surveys mentioned previously revealed that 28 per cent of the population would consider 'park and ride' instead of driving all the distance of their trip if they were holders of such a transport card.

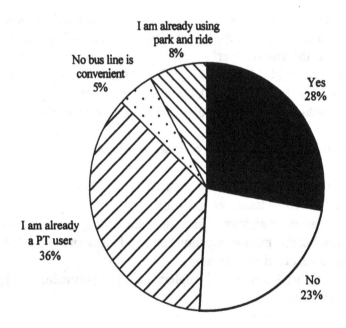

Figure 11.2   If electronic payment was implemented in all bus lines and all parking areas of Thessaloniki, would you then prefer to park and ride instead of using your car for the whole trip?

Source: ADEPT II.

Increased Market Penetration

In the framework of DISTINCT, further functionalities, mostly non-transport, have been added to the same card, namely access to health records, access to information, as well as specific services for the disabled, such as access to off-road parking spaces. The migration to a multi-function smart card, which accommodates cross-sector integrated applications, is accelerating and widening the market penetration of smart cards, and furthermore, is potentially increasing the use of public transport. These conclusions are supported by further end user surveys conducted in the field in summer 1999 and in January 2000. More particularly, the willingness to acquire a smart card increased from 63 per cent of the general population in the case of a simple transport payment card to 81 per cent for a multi-function DISTINCT card.

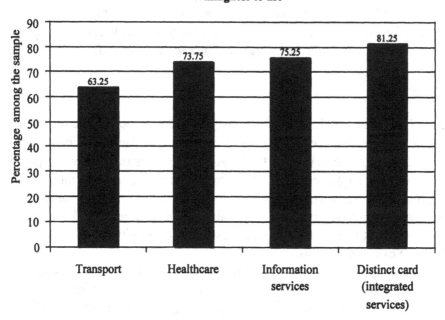

Figure 11.3   Public's willingness to acquire smart cards for various purposes

Source: DISTINCT.

Furthermore, while 51 per cent of current DISTINCT cardholders acquired the card primarily for other services than public transport payment, 87 per cent stated that they expected to use their card for all services, including public transport. The conclusion is that a multi-application smart card can introduce people to the merits of public transport, and thus increase patronage among non-users of the public transport system.

Financial Analysis Results

Financial benefits are expected to occur both because new users will shift from other modes and because of fraud reduction (which at the time of writing is around 15 per cent), as well as thanks to the decrease in ticketing management costs. Financial analysis is concerned only with the marginal financial impacts of the investment (application) under examination. Hence only marginal (with versus without the system) economic impacts are being computed.

By taking into account financial benefits as well as capital and additional maintenance costs, a financial internal rate of return (IRR) equal to 38 per cent is evident. The results indicate high financial viability for the investment (application). The analysis has proved that for the overall multi-application scheme the pay-back period would be in the range of three to four years, given that initial assessment assumptions were indeed materialised.

Drawbacks

One of the main conclusions of the analysis of the service providers' survey results has been that *user-organisations have not yet adopted the information age culture* required to migrate to electronic, paperless transactions imposed by smart card implementation. This situation is further complicated by the existing legal framework, which made some of the services, namely the parking payment, virtually non-applicable. Indeed, Greek law did not take into account at that time (1998), the possibility of parking payment without the issue of paper receipts.

With respect to multi-application, the major problem has been the *organisational aspects* of such a scheme. The main objective which underlines all projects under consideration is the use of the same smart card for all the systems and, hence, the integration of the operation process for the three transport services and the remaining non-transport ones. The impact produced by the services' integration of the entities concerned is much more severe than what would have been produced if an isolated application was implemented. An integrated smart card system undoubtedly requires the generation and association, for example, of common payment functions, which formulate the site specific *payment model*, upon the *agreement* of the entities involved.

Finally, the *privacy issue* is important, related to the data control and access, and security as well as transparency of transactions. The Greek Data Protection Act requires the protection of all sensitive and confidential information held by individuals and/or commercial entities. In addition, respondents have strongly stressed during the user behaviour 'before' surveys, that the system should ensure their privacy. Only 37 per cent of the sample would be willing to provide their personal data to the card issuers *unconditionally*. This figure is not significantly different (statistically speaking) than the same figures concerning individual services (36 per cent for transport services, 35 per cent for information services), with the exception of the health care application; unsurprisingly, the corresponding figure for health care is 46 per cent. Indeed, users feel more comfortable in sharing their personal information for medical purposes.

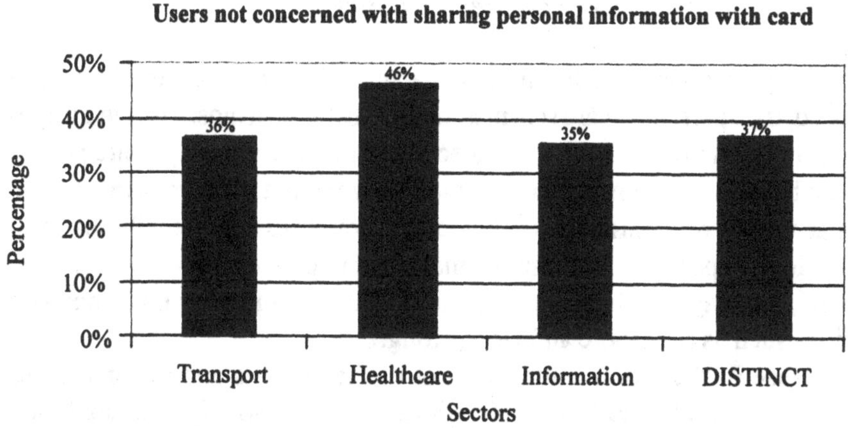

Users not concerned with sharing personal information with card

Figure 11.4   Public's willingness to provide their personal information

Source: DISTINCT.

Conclusions

The Thessaloniki Experience

The Thessaloniki site's work has proved the technical feasibility of the pilot attempts and provided valuable input data for future technology implementation plans, as well as a sound analysis of the Greek market. Most of the evaluation carried out was, however, based on a limited number of test cases or an ex-ante approach (for example, financial analysis). Pilot evaluation results have not been verified yet at full scale.

The Thessaloniki project has, nevertheless, paved the way from traditional fare/toll collection systems of the past to the automated, electronic ones. The

National Highway Fund (operator of the tolls along PATHE Highway) planned to incorporate the DISTINCT guidelines for the future implementation of a smart card payment system. Additionally, fully automated, transponder-based, electronic fee collection is being planned. The public transport operator of Thessaloniki buses is taking under consideration the option of introducing smart cards for e-ticketing within the next few years. Greek municipalities are also discussing smart cards as a cost efficient alternative to coin/token payment of parking services on and off road.

However, a multi-application smart card scheme is still a non-issue.

The Experience from More Advanced Countries

As already mentioned earlier in this chapter, other parts of the globe have initiated, transport smart card schemes, or plan to do so, in most cases in a multi-application environment. These projects, are however, mostly quite recent, but are based on previous extended research work and pilot implementation, similar to that of Thessaloniki. Only a few countries can actually claim to have a long experience in transport smart card implementation, for example, Finland, Norway, The Netherlands, but most importantly Far East countries or cities such as Singapore and Hong Kong.

Because of the recent nature of most of the projects and in some cases because of confidentiality considerations, there no widely available data concerning the real performance of smart card scheme operations.

Figure 11.5 illustrates how migration from a single public transport smart card to a multi-application one has resulted in increased profit for the Connexxion company (a bus operator in the Netherlands).

Transport Smart Cards in Southeastern Europe

Most countries of Southeastern Europe are less advanced in socioeconomic terms than Greece and, of course, Western Europe. Therefore, conditions seem less favourable for the deployment of smart card systems in this area, at least at the moment, especially if the drawbacks mentioned above are considered. The experience of other countries, and most importantly of Thessaloniki, which has similar characteristics to other Southeastern European cities, could be a helpful guide for the smooth introduction of smart card schemes.

This is generally true for most countries of Southeastern Europe. Nevertheless, there are some significant exceptions. The most important is Turkey. There are currently two large projects running in Turkey: one in

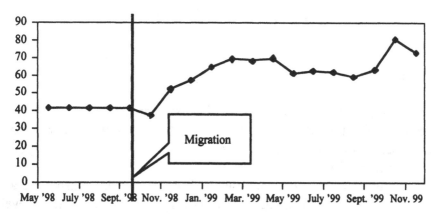

Figure 11.5   Monthly turnover in Euros (x1,000); payment by smart cards in Zeeland on 220 buses

Source: DISTINCT and Connexxion.

Istanbul (buses will be equipped with a contactless e-ticketing system) and one in Izmir (multi-application transport card for all public transport modes).

The Interoperability Stake

One of the most strongly debated issues at both national and international level is that of the transport smart card's interoperability, and it is perceived as the crux of the matter within the smart card and transport community alike. Interoperability applies at the following levels:

- local/regional level, for example, interoperable e-ticketing between buses and trams or metro;
- national level, for example, interoperable e-ticketing between local urban public transport and inter-urban railway or toll collection;
- international level, for example, interoperable e-ticketing across borders;
- transport and non-transport services: generic multi-application smart card;
- payment means level, for example, interoperability between e-purse and e-ticket payment means.

The following measures and actions will ensure interoperability:

- finalisation of standards (either ISO or CEN level);
- national initiatives.

International Initiatives

There is significant ongoing work to create standards, most of which have already provided results, for transport applications in order to permit smart card-based multimodal transport journeys that may well span local, regional and national borders.

At national level, most countries with a tradition in transport smart cards have established committees and open fora to produce national level guidelines for interoperability, for example, Germany, France and the UK. A 'good practice' example is that of the UK and ITSO (Interoperable Transport Smartcard Operations) supported by the National Department of Transport. ITSO has published national standards for the accomplishment of a nationwide interoperable smart card public transport ticketing scheme.

At European level, the long experience in the research field of smart cards for transport is crystallising within the efforts of the e-Europe initiative. More particularly, TrailBlazer 9 of e-Europe is dealing with smart card-based payment of transport and has the following objectives:

- to promote the consistent use of smart card technology in public transport in corporation with 'associated' economic sectors (i.e., parking, tolling, leisure, sport, culture, etc.);
- to ensure and push forward interoperability (i.e., seamless travel) between different ticketing systems;
- to enforce efforts to build and create reliable business cases.

The results of the transport TrailBlazer 9 of e-Europe will undergo a standardisation process through the conduction of a CEN/ISSS workshop, titled FASTEST (Facilitating Smart Card Technology for Electronic Ticketing and Seamless Travel). FASTEST results are available at http://www.cenorm/be.

References

ADEPT II (1998): 'Report on the Installation and Evaluation at the Thessaloniki site', D5.2 (public deliverable), <http://www.cordis.lu/telematics/tap_transport/research/projects/adept2.html>.

DISTINCT (1999a), 'Exploitation Plan: Thessaloniki Site', D7.2.3 (public deliverable), <hermes.civil.auth.gr/distinct/>.

DISTINCT (1999b), 'Description of the Thessaloniki Site', D10.1 (public deliverable), <hermes.civil.auth.gr/distinct/>.

DISTINCT (1999c), 'Integration of the DISTINCT functions in the Thessaloniki site', D10.2 (public deliverable), <hermes.civil.auth.gr/distinct/>.

DISTINCT(2000a), 'Business Plan for the Exploitation of DISTINCT', D7.2 (public deliverable), <hermes.civil.auth.gr/distinct/>.

DISTINCT (2000b), 'Results of the Evaluation and Validation of the Thessaloniki site', D10.3 <hermes.civil.auth.gr/distinct/>.

Frost and Sullivan (1999), 'Worldwide Smart Cards Application Markets – Introduction, Executive', <http://www.marketresearch.com/map/prod/923528.html>.

Kyrou, G., Nikolaou, K., Toskas, G. and Tsilikas, N. (1998), 'An Integrated Plan Including Transportation Planning, Environmental Policy and Telematics Applications for the Metropolitan Area of Thessaloniki-Greece', proceedings of the *World Automotive Congress FISITA 1998*, 27 September–1 October, Paris.

Mintel (1998), *Smart Cards*, Mintel International Group Limited, London

Mizaras, V. and Aifadopoulou, G. (2000), 'Multi-application Smart Cards Stimulate Public Transport Usage', paper delivered at the *Transport and New Technologies* conference, 30 November–2 December 2000, Piraeus.

Mustafa, M., Giannopoulos, G., Nikolaou, K. and Toskas, G. (1999), 'Using Smart Cards for Integrated Telematics Services: The DISTINCT Approach in Thessaloniki', proceedings of the *6th World Congress on Intelligent Transport Systems – ITS '99*, 8–12 November, Toronto.

Nikolaou, K., Toskas, G., Mizaras, V. and Basbas, S. (2000), 'The Environmental Aspect of Integrated Smart Card Based Services in Thessaloniki', *Fresenius Environmental Bulletin.*

Nikolaou, K., Toskas, G., Mizaras, V. and Basbas, S. (in press), 'Planning of Telematics Applications for Transport and Environment in an Urban Area', *Journal of Environmental Protection and Ecology.*

Chapter 12

Tools of Market-oriented Rail Restructuring with Special Regard to Improving Management Information Systems

Zoltán Bokor and Katalin Tánczos

Introduction

The results of long-term transport performance and market share analyses show that the role of railway transport has significantly decreased in the past 25 years. This has been caused by the structural and behavioural changes in the economy and the inflexible operation of railway companies. According to the results of comparative analyses, the observed development tendencies are differentiated by the technology level and operation circumstances of railways.

Practical experiences verify that market orientation in the framework of appropriate regulation can be a successful means to consolidate business processes. This is more and more important in the case of today's railway companies.

The chapter aims to discuss the following points:

- identifying a set of tools which make the operation and management of rail companies more effective;
- analysing the strategic planning process of rail companies which helps to find a suitable model for adapting the identified tools;
- selecting the control-based cost and revenue management for a detailed analysis and evaluating models for applying the methodology for the case of rail transport;
- evaluating a control-based management system model for rail companies;
- describing the operation processes of the rail controlling information and management system model;

- making recommendations for the practical adaptation of the evaluated models.

The Main Tools of Market-oriented Rail Transport

The means of rail market orientation collected on the basis of national and international special literature can be systematised according to defined factor groups. The factors and factor groups are not independent, so the system of the relationships between them has to be explored so as to prepare any practical adaptation. The most important rail restructuring factors, including their cause-effect relationship, are shown in Figure 12.1. National and international transport regulation determines the main frames for rail strategic planning. Rail strategic planning results in marketing and innovation policies (in strong interaction). A further important outcome of strategy making is the reorganisation policy of the rail operation and management system. Finally the effective process of business reorganisation (BPR) needs to be supported

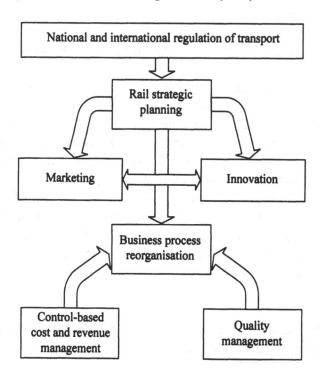

Figure 12.1   Means of market-oriented rail restructuring

by control-based cost and revenue and by implementing quality information/ management systems.

The practical use of the identified means can be realised in a framework of a model being suitable for market position of the given railway company. That is why the core of rail restructuring is the strategic planning process. Using the tools of strategic modelling, possible development images can be set up along specific dimensions. On the basis of selected relevant images, market orientation models for railways can be developed that consist of different combination of the identified factor groups. The models help railway companies to develop a strategy that improves their efficiency and competitiveness.

Figure 12.2 illustrates the strategic modelling process for the case of a rail company. The analysis of the external and internal operation circumstances results in three key performance indicators:

1   intensity and pressure of regulation concerning market orientation – as most important external influencing factor;
2   general state of development (technical equipment and management practices); and
3   range of operation (structure and geographical extension of rail transport services) – as most important internal influencing factors.

Along with these factors – as strategic dimensions – development scenarios for railway companies can be examined. The most relevant scenario identities the strategic goals, which describe the efficient operation form of the given rail transport system. To reach the target position an appropriate restructuring model has to be chosen. The model contains the market orientation tools meeting the requirements of selected strategic directions.

A suitable rail restructuring model involves marketing and innovation policy, as well as a business reorganisation plan. Although all the above-mentioned elements of the restructuring model play a significant role in improving effectiveness of rail systems, this chapter pays special attention to the BPR and cost/revenue management.

Business Process Reorganisation and Cost/Revenue Management by Applying Control-based Modelling Methods

Effective cost management is one of the most important market orientation

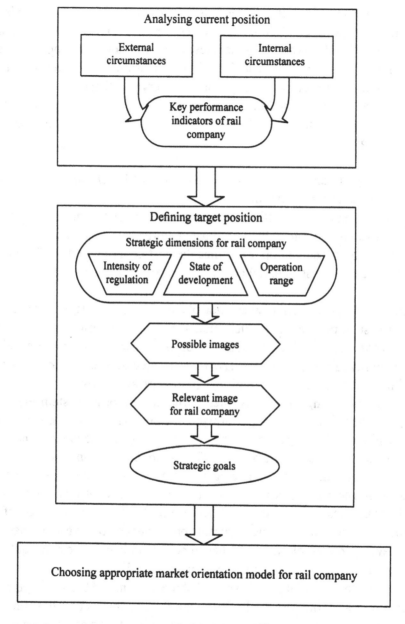

Figure 12.2   Model of rail strategic planning process

tools. Adapting and making practical use of the latest controlling methods can contribute to the success of rail management: an adequately elaborated control system which collects and exploits relevant information makes operative and strategic planning more reliable. By using such information systems even the implementation of reform strategy can be observed efficiently.

The main phases of building an operative control management system are summarised in Figure 12.3. The first task is to identify the tools of operative controlling:

- cost object calculation: examines the costs and performance outputs in business/organisation units;
- internal service calculation: examines the costs of complex services (for instance maintenance) between business/organisation units;
- profit object calculation: examines the cost covering and profit rates of products (for instance, rail passenger or cargo transport tasks).

To prepare the practical implementation of control-based management methods for the case of railway companies, models based on domestic circumstances have to be elaborated. Moreover, an appropriate organisation structure and the acceptability of new procedures have to be ensured. A further task is to assess the informatic supporting possibilities (hardware and software).

When modelling the management structure the basic system elements (objects) are defined, then the performance connections existing between them are identified and finally the data structure and the main structural or functional data concentration areas are built up. The dynamic operation model based on a system model describes the management activities – planning, accounting, analysis and decision-making – using mathematical functions. The continuous business process re-engineering influences the structure and operation of the controlling model, but its basic organising principles do not have to be changed. The developed controlling model has to be able to handle the consequences of structural changes to ensure long term applicability.

A possible controlling system model of the reorganised business operation structure of a market-oriented rail (holding) company can be examined in Figure 12.4. The main business processes according to the new rail operation model are:

- the two business branches operating in the transport market are the rail passenger and cargo transport units. Their products (profit objects): transporting passengers and goods. Producing these products is the task of

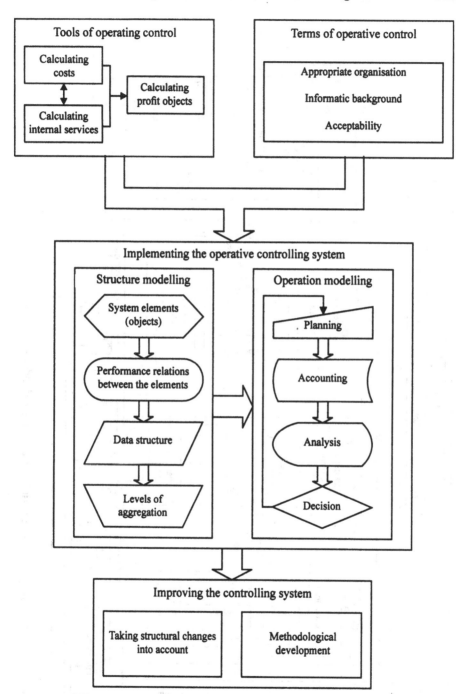

Figure 12.3 Phases of building the operative control management system

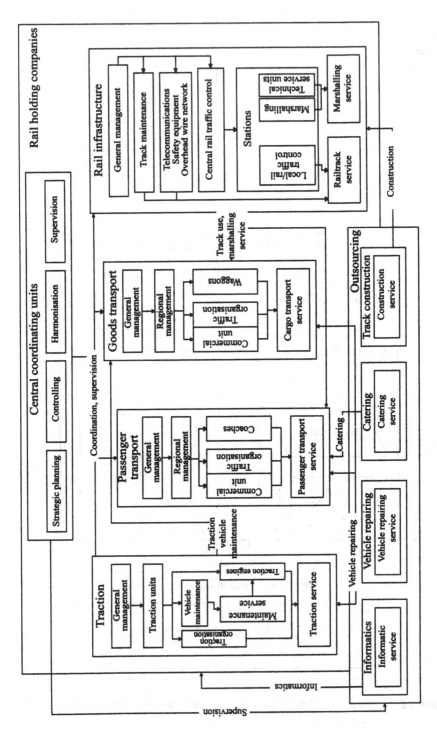

Figure 12.4   Controlling system model for a reorganised rail company

commercial, traffic organisation and vehicle (coach or wagon) management units (cost objects). The two rail transport branches use the services of traction and infrastructure companies.

The infrastructure management branch produces rail track services (profit objects) and provides track (including traffic control) and marshalling to rail transport companies. Infrastructure maintenance units as well as central and local traffic control units (cost objects) both take part in providing these services.

- The traction branch provides traction and vehicle maintenance services to rail transport companies (profit objects). Maintenance is done for engines too. Traction processes are carried out by traction organisation, and engine management units (cost objects). Maintenance projects are produced by vehicle maintenance units (cost objects).
- It is reasonable for a rail holding company to outsource some activities, which do not belong to its main activities. These 'background' units provide such services like information management, vehicle repairing, track construction or catering.
- The central coordination units (cost objects) supervise the operation of the self-managing branches. They define strategic development directions, operate company-wide controlling information systems, coordinate business and technological interactions between the branches and supervise safety.
- It is also important to establish an organisation unit for the fair and effective division of track capacity (not showed on the figure). This organisation unit can operate as a part of the infrastructure branch or the central units. The main point is that it have to divide capacity in an independent way, not giving special preference to any rail transport company. A further task is to develop and use an infrastructure charging system based on the results of (controlling aided) cost covering calculation and the consideration of other economic factors like increasing market share.

The management system of the Hungarian state railway company (MÁV) has been changed significantly since 1994, and it has moved towards the business operation system described before. The main points of restructuring were:

- the state owned company was converted to an independently operating business company, the main shareholder henceforward being the state;
- the accounting systems of rail operation and rail infrastructure were separated;

- the main activity areas operate as partly independent branches within the company;
- integrated transactional information systems for rail technology and financial/accounting management have been implemented;
- regional marketing offices have been set up to increase sales in rail personal and goods transport.

Further restructuring plans are the next issues:

- separating the organisations of rail operation and rail infrastructure;
- establishing a rail track capacity divider unit as a part of the rail infrastructure company;
- improving the rail infrastructure pricing system based on marginal user costs;
- forming a rail holding in the long run.

Operation of Developed Management Information System

The operation phases/activities of control-based management information system are shown in Figure 12.5. The first phase is planning. Planning is based on the so-called meta database, which contains the controlling system models and technological data about activities carrying out in the railway system. The cost, performance and revenue plans are produced according to cost and profit objects, then the plan data are made available to the calculation procedures.

The next step is preparing the actual data for the calculation. After collecting basic data they will be mostly processed to meet the requirements of controlling model standards.

When calculating cost objects the elementary internal services/performances have to be priced, which makes it possible to drive the performance-based operation costs (so-called secondary costs = costs of internal services) between cost objects. Cost objects can be evaluated by collecting primary costs, driving secondary costs and measuring performance. Further task is aggregating data of cost objects to evaluate operation cost and performance efficiency of bigger business/organisation units as well.

The cost side of the profit objects consists of direct product costs, and direct operation costs driven from cost objects (in performance-based way). After adding the revenue side, it will be possible to evaluate direct cost covering of profit objects. The indirect operation costs are included on aggregated levels

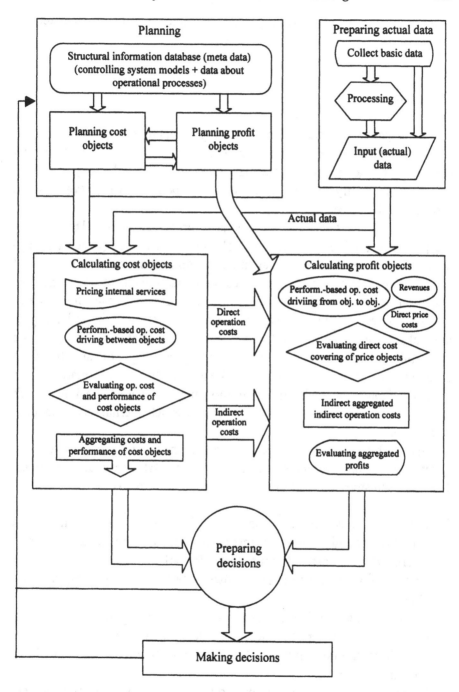

Figure 12.5   Operation model of the control-based management information system

of profit objects, which makes it possible to calculate profits of main activity areas.

Conclusion

The management practices of most railway companies do not yet meet the requirements of market-oriented business organisation. They have mostly centralised management and information systems, which do not pay enough attention to identify the real operation costs, performances, revenues and profits of rail services.

The management of rail companies can be gradually improved by adapting the proposed models and methods. The practical use of (at least partly) decentralised controlling management systems makes the management activities more treatable and improves the establishment of decisions. The operation costs and performances of different structural units (for instance commercial, rail traffic organisation or control units), and the efficiency of internal services (such as maintenance) become analysable. Furthermore the profitability of different products/activities (e.g., passenger or cargo transport tasks) become identifiable exactly. On the basis of former information it can be stated among others what rationalisation measures have to be done, which external service activities should be preferred or which internal services should be replaced.

That is why it is recommended to reorganise the organisation structure and to develop control-based management information systems for rail companies. The proposed steps to build such systems are the following:

1 forming a new organisation structure suitable for decentralised management and at the same time for national specific characteristics;
2 determining the cost and profit objects for rail company;
3 determining the performance flows between the cost and revenue objects;
4 evaluating a controlling system model for rail company (based on the object structure and performance flow determined before);
5 describing the cost and profit calculation processes for controlling operation model;
6 implementing the developed controlling system and operation models by applying appropriate business intelligence tools (= software with controlling functionality);

7   organising data flows and evaluating reporting facilities for control-based rail management system;
8   training the user staff of controlling system.

As a demerit of decentralised rail management systems the increasing coordination demand of the (at least partly) independent operating rail branches can be mentioned. The solution to this problem is an integrated management information system with strong informatic background.

Finally, we must remember that some developing tasks, such as improving interoperability or cost effectiveness, are common to all southeastern European railways. But there are also particular national characteristics, such as customer habits, traditional practices or geographical position, which have to be taken into account when evaluating operation and management models.

References

Bokor, Z. (1999), 'Applying Controlling Management Methods in Transport I–II', *Transport Science Review*, 10 and 12.

Bokor, Z. (2000), 'Elaboration and Practical Use of the Conditions of Market Oriented Railway Transport with Special Regard to the Controlling Based Management System', PhD dissertation, BUTE.

Bokor, Z. (1999), 'Methods for Cost and Revenue Management of Internal Logistic Processes in Transport Companies', *Logistic Review*, 3.

Bokor, Z. (1998), 'Tools of Market Orientation in Railway Transport', *Management Science Review*, 6.

Nash, C., Preston, J., Shires, J. and Wardman, M. (1994), 'Rail Privatisation: The Practice', Working Paper 420, Institute for Transport Studies, University of Leeds.

Preston, J. (1998), *Railway Reform Entrepreneurship: A Tale of Three Countries*, PTRC European Transport Conference, Proceedings of Seminar G.

Tánczos, K. and Bokor, Z. (1998), *Methods for Efficient Development and Maintenance of Transport Infrastructure*, 17th Symposium on Transport Science, TU Dresden.

Tánczos, K. (1999), 'Conditions of Efficient Operation of Integrated Transport Infrastructure', DSc dissertation, Hungarian Academy of Sciences.

Tánczos, K. (1998), 'Conditions of Interoperability in Rail Transport System', Transport Science Review, 3.

Wissenschaftlicher Beirat (1997), 'Bahnstrukturreform in Deutschland – Empfehlungen zur weiteren Entwicklung', *Internationales Verkehrswesen* 12/97.

Chapter 13

Port Competitiveness as a Function of Trans-shipment Development in the Mediterranean: The Case of the Trieste Container Terminal

Marco Mazzarino

Introduction

It is quite well known that port competitiveness strongly depends on maritime geography, that is, on how shipping companies strategically decide to serve different geographic areas with their services. In a sense, ports very often have to decide their policies on the basis of 'what's going on' in the shipping world. We see, for instance, how trans-shipment development in the Mediterranean, as a strategic decision taken by shipping companies (Zohil and Prijon, 1999), is going to change the competitive position of many maritime terminals in that area dramatically. Terminals which are currently called by direct services are likely to become feeder ports and vice versa. There is sometimes very little in this development which can be strictly linked to port competitiveness; rather, port competitiveness turns out to depend on maritime strategies.

In this chapter the effects of such changes are examined with respect to the Trieste container terminal (Molo VII). First, a likely definition of 'market' (current and potential) for a container terminal is introduced and is applied to the Trieste port. Then it is shown how Trieste's competitive position is likely to change as trans-shipment continues to develop in the Mediterranean. Finally, some policy proposals are put forward as to how the competitive position of Trieste could be enhanced.

A Definition of 'Market' for a Maritime Container Terminal

In this chapter a specific concept of port catchment area is adopted. On top of its 'own' efficiency (supply side), it is clear that the competitiveness of a port is primarily determined by the existence of some market areas (generating and/or attracting commercial flows – demand side) that can be captured by means of suitable and efficient port services. The development of commercial flows which can be potentially managed by a terminal mainly depends on three factors:

- the geographic position of the terminal;
- the characteristics of its hinterland;
- the strategies of shipping companies about port choice criteria.

In the main, the first factor should not be viewed nowadays as the fundamental one. It was actually important in the past, although today linking port competitiveness to geographic criteria alone can be at best a good starting point. Instead, the fundamental criteria are those related to the economic 'content' of transport services among a certain number of possible points of origin or destination of flows. In other words, if one wants to estimate the market position of a port one must analyse the economic and logistics characteristics of the transport services which are of interest for the port, both maritime and land-based (rail, road, etc.).

If transport services characteristics (in terms of economic and logistics convenience) among origins (O) and destinations (D) of potential interest for a terminal are used as the overall criteria for evaluating the competitiveness of the terminal itself, we come up with identifying as many markets as the number of O–D pairs finding it economically viable to use the port. Market segmentation then reaches a quite high level: what is of importance is no longer the market 'area', rather, it is the specific O–D pairs that are assessed on the basis of some sound economic criteria. These criteria can be summarised in the concept of generalised cost, which in turn includes a number of aspects such as:

- infrastructural aspects: existence of efficient land connections (rail, road, etc.), quays, etc.;
- operational aspects: transit time, costs and tariffs, safety level, marketing, etc.

In addition to this, political aspects can be considered, such as those concerned with relationships among countries.

If one relates the concept of 'market' to specific O–D pairs (specific in terms of direction, mode choice, etc.) the notion of market 'area' should be reformulated. Loosely speaking, a market 'area' can be seen as an aggregation of some origins and/or destinations which can be located overseas or in the hinterland. Moreover, the geographic criteria can be used in order to find 'reasonable' origins and destinations which can bring about the existence of some markets (in a strict sense). In other words, some origins and destinations can be cut off on the basis of some geographic criteria as they are judged to be unreasonable.

Identification of Markets for the Trieste Container Terminal

Given the above definition we use it in order to identify the market areas (as aggregations of O–Ds) for the Molo VII container terminal, that are:

*   hinterland market areas: Central and Eastern Europe (Austria, Bavaria, Slovenia, Croatia, Czech Republic, Slovakia, Hungary, Yugoslavia) and north–northeast Italy;
*   overseas market areas: Far East, east Mediterranean, North Africa.

In terms of specific O–D pairs, the relevant markets turn out to be:

*   North and North East Italy (specifically Pianura Padana, Veneto and Friuli – Venezia Giulia) – Far East and East Mediterranean. In these markets there is a strong competition from the Tyrrenean ports;
*   Switzerland – Far East, where we again have competition from the Tyrrenean Ports;
*   Bavaria – Far East. We have here competition from northern Europe ports;
*   Austria, Hungary, Slovakia, Czech Republic – Far East. Again, we have here strong competition from northern range ports;
*   Slovenia, Croatia, Albania – Far East. On these markets we have competition from the Greek ports which increases as we go southward in the Balkans.

Besides these main markets there also exist local markets which are not yet very relevant for containerised flows.

In order to have a general picture of containerised flows of Trieste one can look at Figure 13.2 (TEUs). However, the crucial question is how much the above markets count. That is, what is their relative importance right now?

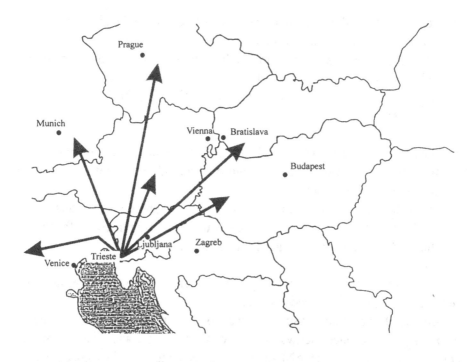

Figure 13.1   Main hinterland links of Trieste

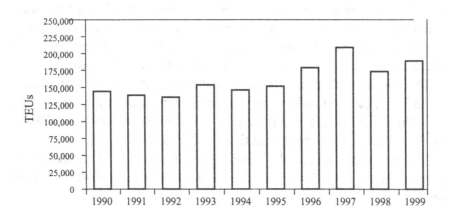

Figure 13.2   Trieste container throughout

In order to answer this question it is crucial to examine the situation of the maritime routes and services which are of relevance to Trieste.

The critical variable for the shipping lines in order to decide the strategies related to port choice (route network) and markets is the traffic flow managed by each port. Currently the main maritime services calling the Trieste container terminal are:

- the Medway line of Evergreen-Lloyd Triestino Consortium, which serves the Far East and Southeast Asia and produces about 60 per cent of total traffic of the Trieste terminal;
- the MSC services to/from Egypt, Israel, Turkey and the Black Sea;
- the Maersk services calling Trieste through feedering in Gioia Tauro;
- the ZIM services to/from the Middle East and Far East.

Of the overall traffic in Trieste, 88 per cent is due to direct services and the remaining 12 per cent to feeder services.

Among all the containerised services, no doubt the most important one is the Evergreen-managed Medway service on the Far East route. The Taiwanese company currently serves the Mediterranean basin with the following network: Suez–Gioia Tauro–Genoa–Marseilles Fos–Barcelona–Valencia–Trieste–Suez. Operationally, after discharging at the Gioia Tauro terminal for the distribution in the East Mediterranean by means of feeder services, Evergreen discharges about 1,500 TEUs each week in the Genoa port in order to serve the market area of north/northwest Italy. Then it serves the French and Spanish ports and comes back northward to Trieste, where it operates the most important volume of loadings in the Mediterranean, i.e., about 1,500 TEUs each week. In Trieste it collects the export flows from north and northeast Italy and Austria.

The Trieste and Genoa ports then refer basically to the same market areas (north Italy), but the first is chosen by the shipping line for the export flows while the second for the import ones. In this way, export flows from Italy to the Far East realise a better transit time (five days fewer).

The current situation for the Trieste terminal then shows a disequilibrium between import flows (which are fairly small) and export flows (which are quite large).

Possible Competitiveness Scenarios

How is this situation going to develop for Trieste as new developments in the

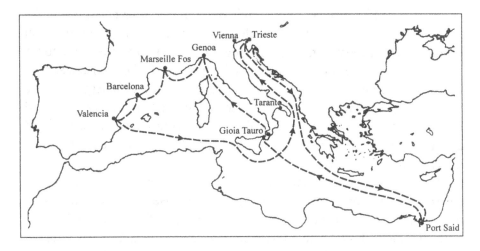

Figure 13.3   Mediterranean ports and Evergreen main network of interest for Trieste

trans-shipment sector are going to occur in the Mediterranean (Sutcliffe and Ratcliffe, 1995)? Trieste is bound to change its competitive position with the 'new entry' in the trans-shipment sector given by the Taranto trans-shipment terminal, which will be managed by Evergreen. As this new terminal enters into operation, three main scenarios can be figured out for Trieste and generally for the north Adriatic terminals:

1   the Mediterranean is served by an ocean-going service calling at Taranto, from where two feeder services are put into operation, one to the north Adriatic (Trieste) and the other to the Tyrrenean, Spanish and French ports. Having served Taranto, the direct service goes back to the Far East;
2   the Mediterranean is served by a direct service calling at Taranto and Trieste, while the Tyrrenean, French and Spanish ports are served by feeder services. After calling at Trieste, the direct service goes back to the Far East;
3   the direct service calls at Taranto, the Tyrrenean and French/Spanish ports, while Trieste is served by feeders.

Clearly, these scenarios are designed according to some economic rules which we discuss later. It is for now quite clear that the 'best' scenario for Trieste is the second one, in which Trieste could be the main port for the shipping line both for incoming and outgoing container flows. Yet we cannot ignore the worst-case scenario, i.e., the third one, in which Trieste is called at by feeder services only. In this sense, it has to be noted that at the time of writing Trieste is slightly preferred to Venice due to some structural characteristics (deep water, absence of fog, etc.) even if Venice is nearer to the same market areas of the Trieste national hinterland. If these characteristics have some meaning for a direct call, they could not be sufficient to choose Trieste as far as feeder services are concerned. In the third scenario it would be possible that not only import but also the export flows would be managed by the Tyrrenean ports and Trieste would certainly run a considerable risk in terms of competitiveness.

It is quite evident that there is an urgent need for Trieste to create the necessary conditions (Hoyle, 1999) so as to realise the second scenario, otherwise it risks also losing its competitive advantage for the export flows. In order to achieve this objective Trieste must act on a critical decision variable: the traffic volumes that it is able to manage in the near future. That is, only if Trieste is able to attract more traffic will it improve its competitiveness. The crucial question is, what kind of traffic and how to capture it?

Traffic volume is indeed one of the most important variables upon which a shipping line decides between two possible strategies of port choice – direct call or feeder service. In fact, a hub and spoke network implies, with respect to a network of direct calls, smaller operational costs for the mother vessel insofar as the ship turns out to be bigger and therefore can better exploit the economies of scale. On the other hand, a hub and spoke network implies additional costs due to:

- trans-shipment operations between mother and feeder vessels, such as terminal costs, inefficiencies, wasted time (and consequently less transit time), etc.;
- the cost of the feeder vessel which clearly can exploit economies of scale marginally.

Generally speaking, how can a shipping company realise the economies of scale due to the mother vessel? It can act on three aspects:

- it can diminish the length of the overall itinerary (network); that is, it can cut the total *distance* by properly choosing ports;
- it can increase the *volume/quantity* of cargo transported: as ship dimensions increase unit operational costs decrease;
- it can save *terminal costs* by choosing convenient, efficient and effective ports.

Among these aspects the importance of the 'volume/quantity' aspect is clearly relevant. Therefore, the more a port is able to attract traffic, the more convenient it if for the shipping company to operate a mother vessel.

Currently, future scenarios for the Trieste terminal are undoubtedly partly at risk. In fact, the bulk of the containerised traffic is strongly unbalanced (say, asymmetric) in that:

- there is only one major shipping company operating the traffic (Evergreen);
- there is only one major overseas market area (Far East);
- there is only one major inland market area, i.e., there exists a strong market share of the national traffic with respect to the (traditional) international one. In particular, Eastern Europe's share is very low;
- there is a strong incidence of export flows.

In addition to this, the situation is certainly going to change rapidly with the development of the trans-shipment in the Mediterranean. The situation must then be improved and a better equilibrium of different factors should be attained.

In order to increase the traffic volumes, the Trieste terminal should adopt a strategy of market segmentation. In this sense, an increase of traffic volumes, above all in terms of import flows, can possibly be derived by the Eastern Europe countries and generally by the international hinterland. In other words, Trieste should recapture its historical hinterland. If Trieste is able to attract more traffic from Eastern Europe and its international hinterland it will be in a position of reaching a 'critical mass' of flows and therefore shipping lines will find it convenient to call at Trieste with a direct service. Perhaps more importantly, such a convenience will produce a 'multiplier effect' of reinforcing the Trieste position with respect also to the national hinterland by capturing additional import flows: in other words, more traffic from the international hinterland plus additional traffic from the national one. This would be clearly the best possible scenario for Trieste.

If Trieste only marginally captures additional traffic, it can be called at, along with the Tyrrenean ports, by feeder services through some Mediterranean hubs (Taranto, Gioia Tauro, etc.). In this scenario Trieste would lose some of its competitiveness as far as export flows are concerned and it could try to acquire additional traffic from its international hinterland.

The last scenario is the worst one: Trieste is not in a position to capture new traffic from its international hinterland and, for several competitive reasons (with respect to other North Adriatic ports), partially loses its competitiveness for export flows. In this scenario an almost total shifting of traffic from north Adriatic to north Tyrrenean ports is likely to occur, even for the export flows: Trieste ends up being called at by small feeder services only.

Some Policy Guidelines

How can Trieste realistically implement the best overall strategy of 'recovering traffic from its international hinterland' (Goss, 1990)?

It is quite clear that it should act on the competitive factors of the hinterland connections, that is, on the generalised cost parameters such as transit time, costs, reliability, safety, etc. These parameters depend on the infrastructural and operational characteristics of different connections from and to the port. In other words, one has first to identify specific markets (i.e., O–D pairs) and then enforce some supply policies so as to improve the relevant parameters of the generalised cost, and consequently make Trieste more competitive. The goal is that of optimising these parameters and the optimisation process must lead to the configuration of new, more effective and convenient transport services. Then, by means of a marketing process and negotiations with transport and commercial operators these services become able to induce new traffic to be 'channelled' to the port.

From the current situation, the following issues on different markets (O–D pairs) can be identified for the Trieste terminal, both from an infrastructural and operational point of view:

- in the road sector, there is the 'historical' bottleneck represented by the Mestre/Venice node, which strongly hinders connections with the national hinterland westbound. In the rail sector, there are a number of inefficiencies on the eastward connections. Generally, accessibility westward and, above all, eastward must be improved;
- road services on the Austria connections are penalised due to the Austrian transport policies, therefore rail services must be improved on such connections;
- it is vital to improve the rail connections to/from the Czech Republic and Slovakia, which currently tend to prefer the northern Europe terminals;
- it is necessary to improve the tariff competitiveness of rail services to/from Hungary;
- an improvement in the rail service to/from Bavaria is required. Some of

these improvements are already realised (for instance, the new rail service Trieste–Munich managed by MSC).

To be fair, a number of policies of this kind are being developed, especially on the north–south markets. If these policies are effective, expansion of the terminal container will be certainly required. Yet this is currently a matter of discussion in so far as different technical options could be put forward, let alone the ongoing coordination and integration with the nearby terminal of Koper (Slovenia) by means of a mixed company. The detailed analysis of different options, and the related socioeconomic and environmental consequences, is beyond the scope of this chapter. However, it can be said that much of such discussions is about:

- strategic decisions around port planning, i.e. how to assign different functions to different port areas;
- the role of the port in the container sector, given the possible magnitude of traffic forecasts.

One basic issue runs as follows. Operating traffic purely as a container 'terminal' is no longer consider a remunerative activity. On the contrary, one of the current challenges by many ports is that of providing value-added logistics services to the cargo in transit, rather than merely moving the technical units from ship to shore and vice versa. If the port decides to offer such services, then specific port areas must be allocated to such activities. Some local operators and policy makers argue, however, that rather than finding new areas for value-added logistics services, the Trieste terminal should first improve its position as pure 'terminal' (i.e., increase container throughput first): when a proper number of TEUs is reached, then value-added logistics services can also be provided. In other words, the short term goal is that of expanding the terminal itself.

To conclude, it seems that the most effective policy tool for Trieste to expand its international hinterland lies in the development and improvement of rail services, especially in the form of new, dedicated services (block trains, etc.) capable of guaranteeing optimal tariffs and transit times. Trieste is not optimally located with respect to western European markets, therefore its best strategy should be that of transferring containers by rail to the inland terminals that are better located. On the other hand, Trieste is well located with respect to Central and Eastern European countries, for which rail could be the best choice of land connections, given the generally poor road network in those countries. Seen in this way, the main problems in organising rail services to/from the international hinterland lie not so much in the infrastructure sector as in the operation and management field. Very (too) often rail services are difficult to operate due interoperability, reliability, tariffs, transit time and also safety reasons, not to lack of capacity on the routes.

Room for such improvements certainly exist, given also the traditional picture of Trieste as a 'rail' port.

Acknowledgements

The basic ideas and reasoning of this chapter arose from consultative work carried out by ISTIEE for the local authority for industrial development (EZIT) regarding a possible location of a logistics and distribution centre (Distripark) in the Trieste area. Moreover, I would like to thank the Evergreen management staff in Trieste for discussing with us fundamental issues such as port choice criteria and shipping strategies and development in the Mediterranean.

References

Goss, R.O. (1990), 'Economic Policies and Seaports: The Diversity of Port Policies', *Maritime Policy and Management*, 17 (3).

Hoyle, B. (1999), 'Port Concentration, Inter-port Competition and Revitalisation: The Case of Mombassa, Kenya', *Maritime Policy and Management*, 26 (2), pp. 161–74.

Lim, S.K. (1996), 'Round-the-world Services: Evergreen and US Lines', *Maritime Policy and Management*, 23 (2), pp. 119–44.

Sutcliffe, P. and Ratcliffe, B. (1995), 'The Battle for the Med Hub Role', *Containerisation International*, 28 (7), pp. 95–9.

Zohil, J. and Prijon, M. (1999), 'The MED Rule: The Interdependence of Container Throughput and Trans-shipment Volumes in the Mediterranean Ports', *Maritime Policy and Management*, 26 (2), pp. 175–93.

Elements of an Intelligent Transport System-based European Freight Architecture: The THEMIS Approach

George Giannopoulos

Introduction

Freight Transport Trends

In terms of the volumes of transport that are likely to materialise in the coming decades, all indications point to the fact that economic, social, organisational and spatial trends are bringing about a highly mobile society in Europe. In such a society, the movement of goods has a special importance and it as has in the past, it will continue to increase in the future.

By some European Union estimates, characteristically used in support of the TEN-T policies (European Commission, 1997), transport demand as a whole is expected to nearly double by 2010 as compared to 1995. Cross-border traffic is expected to grow by 2–3 per cent per year. By 2010 there will be approximately 30 per cent more passenger cars and 20 per cent more trucks in circulation.

The relative share of transport modes in the total inland transport work is a point of interest and ongoing debate. Over the last 20 years or so, policies have failed to halt the 'onslaught' of road transport in dominating freight transport by land in Europe. The current trends show that, in European Union countries, over the last 20 years road transport has increased its share (in total inland ton-kms) from 50 per cent to 70 per cent. These increases have been made to the detriment of rail and inland waterways, the first reduced from 28 per cent to 15 per cent in freight volumes carried, and the second accordingly. These figures do not include short sea shipping, which if added, would change these percentages somewhat, but not the overall picture.

There is very little indication of the magnitude of transports that are effected by use of more than one modes, in the statistics, a fact that reflects their relative low magnitude in the overall inland transport work, today. The actual figures are 'embedded' in the above figures, but a safe estimate would be that intermodal (in the true sense of the word, i.e., as defined in the existing European Union legislation) accounts for a mere 3–5 per cent in freight transport.

The newly (re)stated European Union policy in this respect as expressed in the new White Paper on Transport (European Commission, 2001) aims at holding the freight transport modal split, by the year 2010, at the same levels in terms of the road transport percentage as it was in 1998, while bringing a decline to the share of road transport in both passenger and freight by 2020 to the benefit of intermodal transport. This will be more pronounced over certain major transport corridors which will be properly 'equipped' to offer a convincing alternative to road transport.

Besides the above estimates and prospects concerning the overall magnitude of (freight) transport flows, and their modal split, their geographic distribution is also subject to change. The main reason for the shifts in the geographic distribution of flows, will be the different rates of GDP growth across the different regions of Europe. As the prospects for industrial development of the less developed European regions in the south, southeast are increasing, the relative growth rate in some western European countries is likely to be lower than that of the other regions. As a result, the volume of (freight) transport is expected to develop at much stronger rates along certain corridors in the southeast. These corridors will be of a western to south, southeasterly direction, from central and western Europe towards Russia and the southeast (Balkan peninsula) area.

The Main Lines of European Union Policy for Freight Transport and Related Issues[1]

In the first decade of the third millennium, freight transport in Europe stands at a crossroads of technological development opportunities that will radically change its face, but also faces as yet unresolved institutional and other policy issues that will determine the range and extent of these changes. As the remaining few restrictions are removed and the liberalisation of freight transport within the European Union countries becomes complete, freight transport operation at European level seems to be proceeding at two speeds. One (applicable to European Union countries), is characterised by high organisational efficiency and free from administrative and other restrictions, led by technological solutions that are now already at various stages of development, and one in the remaining countries of Europe, mainly in the east, continuing to enforce restrictions, and lagging behind in technological efficiency. As the European Union is enlarged, these disparities will tend to disappear.

The current policy aim is for a future European inter-urban inland freight transport system with more market induced quality, and which will be:

- more multi-modal;
- 'heavy' user of intelligent transport systems (ITS);
- widely available to small and medium-sized users; and
- more environmentally compatible.

The structure of the freight transport market as regards the types of companies offering services, is expected to be defined by:

- large size and scale 'mega'-carriers or 'network firms' who will be able to offer competitive integrated transport and logistics services to a wide range of end users;
- 'subcontractors', who will survive with direct connections and 'life support' through subcontracting by the mega-carriers;
- 'cooperatives', i.e., small and medium-sized operators that will 'cooperate' in any sense of the word in order to withstand the competition; and finally
- 'specialists', i.e., firms that specialise in certain types of services that cannot be 'mass produced' by the mega-carriers.

The interurban freight transport 'business environment' of the coming decades is expected to be characterised by:

- higher integration of the transport provider into the whole supply logistics chain. Supply chain management will be the higher order level of management into which transport will be integrated as one of a series of other supply chain management functions;
- closer cooperation and 'integration' with the customer. This will be achieved through more intensive use of information and telecommunications technologies in order to support the large amounts of information flows and data that will be needed between firms and spatially diffused customers;
- urban freight transport will also be dominated by developments in the urban ITS and more particularly in the traffic management systems.

At the same time, rural areas will increasingly become destinations of more and more freight transport movements, and will increasingly acquire the needs of urbanised areas as far as the distribution of goods and freight transport services is concerned. The need for improving freight transport services to rural areas in the future will be much stronger than it is today. Any improvements there will materialise alongside improvements to freight transport services for urban and interurban areas.

The 'enabling' factors for the expected changes in the future European freight transport services will be three main developments: the full implementation of the *new ITS technologies and systems*; advances in *logistics and supply chain management* techniques; and further possibilities that will emerge from *convergence*, i.e., the union of telecommunications, information technology, the Internet and consumer electronics, which will give limitless new telecommunications and computing capabilities.

The timescales for the materialisation of the above are not so much the result of the anticipated speed of implementation of the new technologies and systems (which are likely to mature at very high speeds anyway) but also of the time needed for administrative and legal issues that will have to be resolved. In this respect, Europe has an added difficulty as compared with the US. It is the diversity of national interests and policies that are followed by the various countries, as opposed to the independent but much more uniform approach and policies followed by the US.

Therefore, with the advent of the new technological possibilities, some crucial horizontal and other policy issues have to be taken into account:

- establishment of European standards to cover the functioning of the new freight transport systems within the ITS;
- establishment of mechanisms for continuous monitoring of the function of the market and, if necessary, intervening in order to safeguard the interests of the end user;
- solving some outstanding institutional and legal issues that stand in the way of a wider market implementation of new technological systems. Examples of such issues are questions of liability and authentication in EDI, questions of securing privacy and accuracy in electronic booking and payment systems, etc.;
- making sure that the implications to society and social justice are addressed and dealt with; and

- finding ways to bring into the picture the much discussed, in the past, external costs such as the environmental costs associated with freight transport operation of all modes.

The importance of the above policy issues cannot be underestimated. Past experience teaches us that achieving consensus and political agreement is perhaps the most difficult and time-consuming part of implementing policies for freight transport.

Information and Communication Technologies for Freight Transport

In a demand driven world, the issue of integration between the various parties of supplier/buyer and transport business transactions – i.e., all the actors in a complete freight transport chain – is constantly evolving and is being tackled by a variety of ITS technologies.

Many of the systems concerning the intelligent transport systems aim at making intermodal freight transport chains more attractive by the use of advanced *information and communication technologies (ICT)* applications for transhipment, storage, and transportation.

The key factor in establishing truly intelligent transport systems is to link the different ICT systems and applications together. The large transport companies so far have managed to organise their own door-to-door information systems, because they cover the entire transport chain themselves. Smaller companies, still the majority in the transport business, however, cannot make use of modern ICT because it is not cost effective for them. They depend instead on cooperation with other companies.

Within the overall scope of future freight ITS in Europe as expressed in the previous sections, the ICT that are likely to play the most prominent role in developing new systems and services are shown in Table 14.1.

In the left column of this table there are the technologies that are currently (more or less) available and used in a number of applications, though for some of them commercial systems are yet to be developed. For some of the technologies in Table 14.1, especially the ones expected in the future, some explanations are given below.

- *Distributed database management systems (DBMS)*: the management of the growing linkage of data files located at different sites on computer networks is already a key area for technology development and is expected

Table 14.1   Key enabling technologies (ICT) for use in freight transport

Currently available	Likely in the future
• Data communication protocols and systems e.g. EDI, WWW, MDC, GSM, EPOS and other • Positioning techniques and systems for tracking and tracing (fixed/mobile, terrestrial, e.g., GSM, and satellite: Eutelsat/Euteltracs, Orbcomm, GPS, UMTS-S, Galileo) • Smart cards and tags for remote ID and read/write. • Wireless data transmission systems like the wireless application protocol (WAP) • Secure ID authentication • Electronic and other micro-payment systems	• Distributed database management systems (DBMS) and other data mining and warehousing technologies • Micro and nano-electronics • Further advances in trust and confidence enabling tools • Seamless interoperability of communication networks (multi-domain network management) • Multilingual dialogue communication modes • Embedded intelligence • Middleware and distributed systems • Converging core and access (telecommunications) networks • Data capture and sensors/actuators

to explode in the future. Maintaining consistency is already a challenge. The future ambient intelligence (AmI) systems (see future technologies identified as 'key enabling technologies', in ISTAG, 2001) will bring about DBMS that update dynamically with many distributed devices delivering and accessing data. Currently there are many different types of DBMSs, ranging from small systems that run on personal computers to huge systems that run on mainframes, with many different structures (relational, network, flat, and hierarchical).

- *Data mining and warehousing techniques*: technologies employing further developments in interdisciplinary fields of knowledge encompassing statistics, mathematics, and machine learning to discover relationships that may be present in data in databases.
- *Secure ID authentication, micropayment systems and biometrics*: 'trust technologies' and advanced encryption techniques are strong requirements for both the dependability and the likely acceptance of nearly all of the processes, products and services associated with freight transport. In computer security, biometrics will be important as a means of authentication based on measurable physical characteristics that can easily be checked (fingerprints, iris scanning or speech). These technologies will be used for

access authorisation to areas such as container terminals, customs areas, etc.

- *Micro-payment systems*: practical and widespread use of micro-payment will facilitate freight transport in paying for tolls and other services without time consuming stops. But there is also the possibility that many of the transactions that are on a free basis today will be on a subscription basis in the future.

- *Seamless interoperability* (communication technologies/network management): the technologies that enable complex heterogeneous networks to function and to communicate in a seamless and interoperable way. The focus is on active and dynamically reconfigurable network technologies, methods and tools. Today's concept of hard-wired networks may even disappear, as devices will be established on an *ad hoc* basis using short-range wireless links according to real-time needs. The network software and agent technologies needed to achieve this are not yet available (especially when scale and reliability issues are considered).

- *Data capture and sensors/actuators*: multi-sensor/actuator development and optimisation will enable fast data transfer and identification (e.g., data transfer about wagons' contents and ID while moving).

- *Nano-technology*: is a baseline technology, which will permit the further miniaturisation of devices to (eventually including computer chips) that are thousands of times smaller than current devices. Nano-devices would yield lower power consumption, higher operation speeds, and high ubiquity.

- *Smart materials*: through advances in nano-technology, materials used for various applications (e.g., smart clothing, or seats, etc.) will permit the 'augmentation' of objects so that they can change their characteristics and/or performance by stand-alone intelligence or by networked interaction. Use of these materials will be used to create the 'intelligent' truck of the future.

The flux of the above new technologies (characteristically shown in Table 14.1 and briefly explained above) will form part of the new *ambient intelligence environment (AmI)* that is expected to be in place as soon as 2010 and will be the future transformation of the current technological environment (Ducatel et al., 2001).

The Suggested Approach to an ITS-based Freight Transport Architecture

Attempts So Far

The first attempts at setting up European freight transport architecture, were made by the CEN TC 278 Group 2 (CEN, 2000) dedicated to freight and fleet management. After the design of a first high level reference architecture, this Working Group has been kept dormant for months, waiting for the clarification of a European system architecture context in the area of freight and fleet management and possible extensions to intermodal transport.

Subsequently, two projects – KAREN (1999) and COMETA (1999) – produced two complementary systems architectures as well as relevant standardisation proposals but both of these were focused primarily on general traffic and (passenger) transport systems.

A special 'architecture' workshop held in Brussels in 1999,[2] identified a number of practical and legal problems for a uniform European freight architecture, and recommended that interested parties take more part in, and ensure the coherent ongoing standardisation of, freight transport IT. In the debate it became clear that a number of public and private research projects had been performed without reference to the ongoing standardisation process.

With due regard to a European Commission report (European Commission, 1997) advocating a systems approach to the development of freight transport, and in recognition of the integrated supply chains that are increasingly enabled through transparent IT (information technology) systems, it was proposed that a special forum/workshop be convened to establish just how a fully integrated domestic set of transport standards applicable to continental Europe (including short sea), based on intermodality and flexible modularity, might be developed which could be applied to fully utilise the capacity of freight corridors in all nation states.

This workshop was held in April 2000, almost concurrently with the start of project THEMIS.

The Major Areas of Freight Transport Architecture

In developing a *system architecture,* that can give a plain and functional framework for the development, building and operation of an overall system it is very important to look at the system both from the users' and the operating organisations' points of view. Figure 14.1 shows the different areas where commonly accepted and established freight systems architecture must be developed. This is consistent with the previous attempts made mainly by the KAREN and COMETA projects and the standardisation efforts of CEN TC 278 WG2.

The *transport network management* area includes services and functions related to the transport network (roads, fairways, railroads, lanes, associated equipment, etc.). This includes:

- the establishment and maintenance of information about the transport network or information that might affect the transport network;
- maintenance and control of the transport network and associated equipment;
- traffic flow monitoring and management;
- electronic payment services;
- safety and emergency facilities and services;
- incident management;
- law enforcement with respect to the transport network and the use of the transport network;
- various services such as information services offered to the users of the transport network.

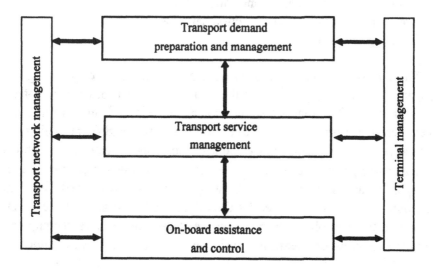

Figure 14.1 Freight system areas of interest in developing a freight architecture

The *terminal management* area is services and functions related to terminals where goods and passengers are transferred between different transport means. The transport means may represent different transport modes. These include:

- information and transaction handling (reception, processing, storage, reporting, invoice handling, etc.);
- multimodal transport synchronisation;
- terminal services;
- resource allocation (also storage capacity).

The *transport demand preparation and management* area is the preparation and management of the transport demands for both freight transport and public transport. This involves pre-trip planning, transport order initiation and follow-up. The services are traditionally offered by a transport agency (freight operator/travel agency). However, one should be aware of the possibility that new services may support the customer in such a way that self-service is becoming more likely. This area includes:

- transport operation preparation (information acquisition, requirement definition, etc.);
- transport operation planning;
- transport operation control and follow-up;
- transport order business transaction handling;
- handling of changes.

The *transport service management* area includes the planning and preparation of fleet operations and the fleet management for both freight and public transport. The transport service management is performed from a static fleet operations centre. This centre can accept goods (freight) or travellers for transportation from one location to another. Routes and timetables are planned. This includes predefined routes as well as dynamic planning depending on transport demands to enable the most optimum transport scheduling. The transport operations performed by the drivers and the transport means are monitored and controlled. The maintenance of transport means and equipment on-board the transport means is coordinated. This function area includes:

- fleet management business transactions;
- fleet operation planning and preparation;

- fleet resource management (drivers, transport means);
- fleet operation control;
- fleet monitoring.

Finally, the *on-board assistance and control* area includes the functions that support the transport of freight or passengers on-board the transport means. On-board equipment should communicate information to the driver and support the driver with respect to the fulfilment of the transport operation. The driver and the operation of the transport means are monitored. The function area includes:

- support for route planning;
- support for pick-up and delivery of freight and passengers;
- transport operation support by the means of information and communication technology (route guidance, traffic information reception and handling, driver control and assistance, equipment control, etc.);
- incident handling;
- after theft services.

The above THEMIS-related freight transport 'architecture' elements are fully in line with previous attempts at creating European freight transport architecture. This refers primarily to projects KAREN and COMETA, and the intention is to build upon their work rather than replace it. Furthermore, THEMIS is in full cooperation and alignment with Working Group 2 of the CEN/TC 278, Committee for standardisation work and architecture formation in freight transport.[3]

Integrating Freight Transport Management and Traffic Management Systems

Project THEMIS's approach to developing an overall freight architecture, concentrated first on the interaction and integration between *traffic* management and *freight transport* management systems.

Such integration can be seen through the axes shown in Figure 14.2. As shown in this figure, apart from the integration of the various systems within the same transport chain (door-to-door), or the same transport operator, the integration of modal *traffic management systems (TMS)* between themselves and with the various *freight transport management systems (FTMS)* deserves particular attention. The idea is that by substantially improving freight transport planning – by using traffic information and vice versa – the quality of traffic information could be improved, with information coming from freight transport management systems.

Integration of TMS and FTMS can generate substantial savings. If a truck could communicate with a traffic information system in the country it is travelling to, it could find out if and at what time the traffic gets jammed around the destination city. This could save time and money, since the driver might decide to deliver or pick up cargo in another city first.

Traffic information can be extremely useful to road freight transport companies. Two types can be discerned:

- historical information derived from readily available statistics and used to feed into the planning system at the home base. Thus, the planning system will take account of the fact that, for example, at 9.00 a.m. traffic is at its worst on a specific highway section.
- real time information made available both at the home base and to the driver. Thus, both the planner and the driver can immediately adapt their trip planning according to the latest information on incidents, roadblocks or weather conditions, or inform the client about delays in delivery.

At the time of writing, road traffic information in Europe is still very much aimed at passenger cars. Few specific applications for freight transport have been developed. There is a large potential market for such information, as many road freight companies might want to buy traffic information if it turns out that they can gain precious time with it. It is not easy to present such a huge amount of information to potential users in a proper format, but technical solutions may be at hand in the near future

When considering the European situation, another complicating factor is added to the picture: the need for cooperation to realise a common communication standard. In Europe, even in the borderless European Union countries, freight transport is international almost by definition. This creates additional challenges, since a large number of parties with different nationalities is involved and there is a lack of standards. Traffic centres using different systems and languages will have to communicate, efficiently and effectively.

There is a need for integration of systems in modes other than road transport. In air transport, extensive traffic management systems have been developed for safety purposes and this information could be useful for freight transport management as well.

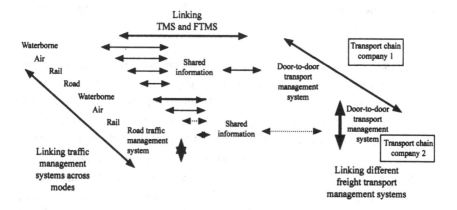

Figure 14.2   The different axes needed for freight transport integration

In the *maritime* sector, a number of initiatives are being developed. An example is a warning system for lockmasters in inland waterway transport. The lockmaster is informed about the approach of a ship with dangerous goods. If the lock is already full, the captain could be told to slow down, resulting in an energy benefit. Or the lockmaster can plan ahead to prevent two ships with dangerous goods to end up next to each other in the locks.

In *rail* transport, the current situation is that the position of a client's cargo is known until it passes the border. After that, a client can do nothing but sit and wait for a confirmation by fax or phone of arrival. It often happens that trucks are waiting in vain at the end of the line, something that is unexpected given the current state of technological developments.

The Prospects for an ICT-based Freight Transport Operation in Southeastern European Countries

By use of the above technologies, and the advent of advanced software development techniques, a plethora of ITS applications are being developed for freight transport. Project THEMIS has surveyed a great number of these (through a survey of other relevant European Union funded projects, and others).[4] We see already a number of these applications in commercial application in western European countries.

The question that arises now, is which of those are likely to be suitable and / or applicable in countries of southeastern Europe, and when. This is discussed in the following section, together with an attempt to categorise these applications so that they are more easily perceived:

1   *Intermodal real time electronic information and transaction systems.* These are electronic and management tools in order to better handle freight transport information especially when operating in a multimodal environment. They include links to ITS-based systems for monitoring and managing vehicles and loading units, at terminals and along the route. These are the most likely systems to be implemented and in the (rather) short to medium term, mainly because they are 'soft' and IT oriented. They are therefore relatively easier to implement without the need for costly infrastructure and they can 'ride' on the IT revolution, which is already beginning to grasp the economies of the southeastern Europe area.

2   *Integrated electronic freight planners* for planning the total (freight) transport chain, load matching, modal selection and combination assistance, and for making the electronic market place available (or 'visible') at a uni- or multimodal context. This is a technology that requires a lot of initial data upon which to build its techniques and functions. These data require a large scale IT infrastructure to be in place and this is not likely to be available in the short or medium run in southeastern European countries. Also the load-matching element has not been greatly popular among Transport operators in western Europe and it is not likely to be so in eastern Europe either. Many questions of data confidentiality and protection from competition have made load-matching systems 'unpopular'.

3   *Document transfer and handling systems.* These are applications that aim at providing electronic completion and transfer of the necessary documents that accompany a load during its transport. Further harmonisation of message and document standards by use of Internet technologies will be necessary before this facility meets with widespread use, but the elements are there and the demand too. It is expected that this technology will easily penetrate the Eastern European market in the next few years.

4   Further use and exploitation of the *Internet applications for e-commerce and e-fulfilment* especially in the context of freight transport. These technologies are penetrating already the (south)eastern European markets, and it is expected that the Internet revolution will not delay its application there as well in the next few years too.

5   *Positioning and vehicle/load tracking and tracing.* These are fairly well developed techniques for vehicle and wagon tracking and tracing. They are still far behind in tracking and tracing for individual loads. The necessary

infrastructure is far from existing in southeastern European countries. It is an open question how fast such infrastructure (and the necessary business structures) can be put in place there. If (and when) the European Galileo navigation and positioning system is put in operation it would be safe to assume that such (tracking and tracing) services can be widely applied in countries of southeastern Europe.

6 *Travel time and ETA (estimated time of arrival) prediction systems.* This calls for full integration of the freight transportation planning and management systems with the traffic management and on-line information provision systems. It also calls for development of application of new traffic congestion prediction techniques and algorithms that will allow timely and accurate information to be given to drivers on the tracks about their estimated time of arrival according to the *prevailing* traffic conditions. Such integration and prediction algorithms are still to be achieved in western European countries although advanced traffic control and management systems are installed and substantial investment is directed towards integration (as part of creating the European ITS). In Eastern European countries, especially souteastern ones, these applications may still be a long way off. Traffic management systems are old and on-line information provision almost non-existent. The application of the above techniques would require extensive investment in new systems and their integration within an overall ITS, something that does not seem to be the priority of the governments in the area for a long time to come.

7 *Order processing and payment.* These include handling of all the necessary documentation for the order and payment processing electronically before and after the trip. This will connect the (end) customer with the (transport) operator for all offer, order, contract, and payment during the whole transport process. Systems like these can be (and some are already) in application in southeastern European countries in the very near future.

Conclusions

Freight transport is a major 'user' of ICT and is expected to be one of the first beneficiaries of the future European ITS. Systems and techniques that are being developed to enhance Freight Transport operation make extensive use of modern *information and communication technologies* and are likely to benefit even further from these in the future. The flux of these new technologies (characteristically shown in Table 14.1) will form part of the new ambient intelligence environment

(AmI) that is expected to be in place as soon as 2010 and will be the future transformation of the current technological environment.

The various applications and systems that are being developed for Freight Transport by use of new ICT and other technological applications, need now to be integrated within an overall ITS-based freight system architecture. Such development will enable further integration and interoperability.

Project THEMIS is monitoring all these developments towards a user-desired end, and develops a system architecture that will be necessary in order to fully integrate freight transport within the emerging European ITS. The elements of such architecture are developed for:

*   transport network management;
*   terminal management;
*   transport service management;
*   on-board assistance and control; and
*   transport demand preparation and management.

Priority is given to enabling the operational integration between traffic and freight transport management systems, i.e., in developing the system architecture on which the existing systems of advanced traffic management in urban and interurban areas will provide the necessary inputs, on-line, for the operation of freight transport management and the estimation of accurate travel times and route finding algorithms.

There is a lot of scope in creating such an overall freight architecture. The primary beneficiary will be the European end-user of freight transport services but many benefits will also result for the operators and wider society as whole. It is hoped that THEMIS's approach, having the support of the European Commission, will provide a much needed catalyst for concrete and conclusive developments in this field in the near future.

Notes

1   See Giannopoulos, 2000.
2   Organised by the CEN TC278, WG2
3   CEN/TC 278 *Road Transport and Traffic Telematics* was established in 1991. Its scope was defined as follows:

'Standardisation in the field of telematics to be applied to road traffic and transport, including those elements that need technical harmonisation for intermodal operation in the case of other means of transport. It supports:

vehicle, container, swap body and goods wagon identification;
communication between vehicles and road infrastructure;
communication between vehicles;
in-vehicle human machines interfacing as far as telematics is concerned;
traffic and parking management;
user fee collection;
public transport management;
user information.'

4   See THEMIS deliverable for Task 3.1, 'Review of System Architecture Initiatives', <http://www.tfk-hamburg.com/themis/>.

References

CEN (2000), 'Freight and Fleet Management Systems (FFMS). Part 1: High-level Architecture Group', CEN/TC278/WG2, Technical Committee 278 Road Transport and Traffic Telematics.

COMETA (1999), 'Commercial Vehicle Electronic and Telematic Architecture', European Union DG XIII (INFSO) project, Advanced Transport Telematics, 4th Framework Programme, Brussels.

Ducatel, K., Bogdanowicz, M., Scapolo, F., Leijten, J. and Burgelman, J.-C. (2001), 'Scenarios for Ambient Intelligence in 2010', Final Report, January, IPTS, Seville.

European Commission (1997), 'Communication on Intermodal Freight Transport', Report No. COM(97)243.

European Commission (2001), 'European Transport Policy for 2010: Time to Decide', White Paper Report No. COM(2001)370, September.

Giannopoulos, G.A. (2000), 'European Inland Freight Transport Scenarios for 2020 and Some Related Policy Implications', ECMT 8th International Symposium on the Theory and Practice of Transport Economics, Thessaloniki, June.

ISTAG (2001), 'Recommendations of the ISTAG, for Work Programme 2000 and Beyond', IST (Information Society Technologies) Advisory Group, E DG INFSO.

KAREN (1999), 'Keystone Architecture Required for European Networks', European Union DG XIII (INFSO) project, Advanced Transport Telematics, 4th Framework Programme, Brussels.

THEMIS (2001), 'Thematic Network in Optimising the Management of Intermodal Transport Services', DG TREN, 2000–2004 (Thematic Network is an initiative of the European Commission's DG Energy and Transport (DG TREN) to promote the integration of Traffic Management Systems with Freight Transport Information Systems).

Chapter 15

Government Policy to Facilitate Private Participation in Tolled Road Infrastructure, through Investment Risk Mitigation Measures: Southeastern Europe and Asian Countries

Gi Seog Kong and Katalin Tánczos

Introduction

The road network is a costly and increasingly scarce resource. The government needs to raise revenue and some part of this revenue should be collected from road users, since to exempt them would be to give them an unreasonable advantage over the rest of the population.

While most countries have historically avoided charging tolls for public roads,[1] many countries have now turned to tolling as a preferred means for financing infrastructure investment. One of the major reasons is budgetary constraint on the public sector. In most countries with toll roads the private sector has been heavily involved in the development of the roads and often thereafter in their operation.

PPI (private participation in infrastructure) toll road projects have been used with ever-increasing frequency and skill around the world in recent years, both in emerging markets and in OECD economies. Developed and developing countries throughout the world have accumulated a diverse base of experience with the building and operating of toll road systems. Over the last decade the demand for high standards increased substantially in East Asian countries such as Japan, South Korea, China, etc. This was a reflection of rapid economic growth and increasing levels of vehicle ownership and use. South Korea, China, The Philippines, Malaysia, Indonesia, Thailand, etc. operate toll roads developed by the concept of PPI. The toll road projects based on PPI have now started to take hold in Central and Eastern Europe as well as

CIS (Commonwealth of Independent States) countries. After the change in the political system (post-1989), gradual revitalisation of the economy and of the fragmented traffic connections began. In this region, governments face dramatic growth of road needs, both for new facilities and for maintenance and rehabilitation of existing facilities. To solve this, Hungary and Poland have used PPI toll roads and Croatia and Romania are also pursuing PPI projects to upgrade or extend their road networks. However, most PPI projects generally need huge initial investment, making continued private funding very difficult. In addition, the projects take up too much time, thus rendering early retrieval difficult. So the implementation of these projects requires both extreme caution and appropriate government support. For these reasons, one of the most important factors in PPI is how to mitigate efficiently the risks that may arise in the course of the road projects. From this perspective, we analyse PPI cases in the relevant regions and suggest how to facilitate PPI in the toll road projects through investment risk mitigation measures by successful government policy.

Main Risks and their Allocation for PPI Tolled Roads

A PPI toll road concession contract is drawn up with the aim of allocating the many risks and regulations of each party's relationship over a long period. The definition of these risks and their clear allocation between the government (public sector; the concession-awarding party) and the private sector (usually a consortium of corporate sponsors and their lenders and investors) are at the heart of the partnership. The simplified risks associated with a toll road project and possible allocation of these risks are as follows.

Construction Risks

The private sector generally assumes construction risks (completion, quality, cost overrun and construction delay). If substantial changes to the specification of the project are requested by government before or during construction, it is more efficient that government bear these risks.

Operational Risks

The private sector generally assumes operation risks (accident damage, latent technical defects, cost overrun and employee dishonesty). Obviously the risk that facilities and services can be provided throughout the contract term to the agreed output specification will initially rest with the private sector under the concession agreement. Clearly risks, which are wholly with the control of the government, will not generally be transferred to the private sector and their occurrence will therefore entitle the private sector to financial compensation.

Commercial Risks

The risks of traffic shortfall and construction of competing facilities will be determined by the terms specific to the concession and those associated with the project's characteristics. The government generally assumes the risk of price control policy (tariffs) for public interests. The other risks of revenues belong to the private sector if not specified in the contract agreement.

Financial Risks

For the allocation of financial risks (inflation, interest rate and exchange rate), different approaches exist. Where the concession is considered as an ordinary private commercial operation, and the bulk of the financial risk stems from inflation, interest rate and/or exchange rate risk has to be borne by the private sector. Taking into consideration the long duration of the contract and eventual impact of these unforeseeable factors on the revenue to be generated by the project, it is better to reach a binding agreement on certain reference points and forecasts and include price escalation and a fair profit-sharing formula in the contract.

Legal Risks

The legal risks (permits, licences and litigation) must be evaluated with particular care having regard to the fact that one of parties to the contract retains a large discretionary power. Furthermore, in many countries the law is undergoing transformation and the legal framework is not settled. It will therefore be important to consider adopting for contractual purposes other legal systems such as those provided by case study or the common law. The legal risks should be identified and assessed with particular care.

Environmental Risk

The environmental risk is one of the major risks faced in the construction of PPI toll road projects and provision of related services. The private sector normally assumes this risk, but there may be some sharing when changes in environmental regulations require a significant capital investment on the part of the concessionaire or limit its ability to deliver the required availability and quality of service.

Political Risks

The political risks (expropriation, termination, limitation of currency convertibility, etc.) can affect the project at any of its stages. An extreme instance of political risk is expropriation, or severe restrictions on the repatriation of project funds. While political risk tends to be exogenous and largely uncontrollable, if the structure of the project involves a direct or indirect government stake, this could influence actions and mitigate political risk to some extent. Both the government and the private sector generally assume political risks by the terms specific to the concession.

Force Majeure *Risks*

Force majeure risks refer to major events that have a dramatic negative impact on project value. Political events such as war, riots or general strikes, and 'acts of God' such as earthquakes, fires or floods fall into this category. The method for allocating the risk of force majeure risk varies from contract to contract. The *force majeure* relief typically applies only to specific, well-defined events listed in the contract; is available only if contract performance is substantially and adversely affected; applies only to extraordinary events, not normal business risks or insurable events; and the relief is limited to the effects of the *force majeure*.

The Positive and Negative Experience in PPI Tolled Road Projects

Some countries in the area under discussion have adopted private sector concessions as their main approach for designing, building, financing, and operating toll roads.

Hungary

Hungary was the first eastern European nation to have built a privately financed motorway network, the M1/M15 motorway (Budapest–Vienna/ Bratislava). The total project cost ECU329m (US$380m) concession had very little government assistance and the concession contract was signed in April 1993, awarding a 35 year-long concession on the basis of BOT (build, operate and transfer). There was no state guarantee for a certain traffic or cash flow level.

After the M1 was opened in January 1996, traffic and revenues were below projection, because most Hungarian motorists elected to take slower, parallel, not tolled routes.

Furthermore, the Hungarian Automobile Club continued to press to have tolls reduced. Unable to meet its debt payments, in June of 1999 ELMKA(Concession Company) was taken over by the Hungarian government.

The successful development, tendering, negotiation, construction and operation of this motorway project proves that highly complex transactions can be completed, provided the appropriate regulatory and legal framework is in place. But the failure to limit its contribution to land acquisition, apportioning the allocation of the entire traffic linked commercial risks to the private sector (and its acceptance by this latter), became a fatal error.

Yet, despite this difficult situation, Hungary's M5 concession is comparatively healthy. The M5 project (97 km, 35 year concession, BOT) is the Hungarian motorway that will link Budapest and Szeged, Hungary's third largest city. The equity of the concession company is 20 per cent of the overall project cost (FRF2,059m or US$280m) The government contribution is land acquisition, a standby operational subsidy, providing of existing 26.1 km motorway and 30 km half-motorway free of charge but now tolled section. Under the concession contract term, government is to provide the concession company (AKA Rt) with a operational subsidy to the company's cash flow deficiencies, up to an overall cap of HUF 9 billion (1993 value equivalent to ECU85m) over a 13 semester period and average traffic volumes in 1997 reached 97 per cent of initial estimates. The Hungarian government actually compensates the contractor directly for implementing toll discounts.

It seems like a mixed application of 'shadow toll' and 'user's pay' policy. Nevertheless, the drivers using this motorway (97 km) have expressed dissatisfaction with the high toll fees.[2]

South Korea

The new airport motorway, which was opened officially on 21 November 2000, started collecting toll fees from 5 December 2000. This motorway is a new road network (40.2 km, 6–8 lanes) that provides connection between the new airport and Seoul metropolitan area.

The motorway project attracted much interest for being the first infrastructure (SOC, social overhead capital) in Korea to build with private investment by the BTO (build–transfer–operate) method. The construction took 60 months and the operation period (29 November 2000–28 November 2030) is 30 years from the completion of construction.

Nearly 1,476,600m Korean won (US$1,230m) from the private sector was poured into the construction of the 40.2 km-long motorway.

The drivers using this motorway have expressed dissatisfaction with the high toll rates[3] of the new motorway leading to the airport. The airport workers who have to use this motorway (except ferryboats) in particular protested strongly to the concession company against the high toll rates. But the concession company's president insisted: 'It is inevitable that users of the new road have to shoulder high charges when considering the huge amount of money we have invested in this project.'

To solve this problem, the Korean government agreed to compensate the concession company's implementation of toll discounts for airport workers until April 2002. The government's contribution to this motorway is land acquisition and site delivery. The government and concession company share some risks (revenue shortfall, foreign exchange loss, etc.) The government recognises buyout right in the case of defined events listed in the concession contract.

Risk Mitigation Measures

The PPI toll road projects include many components and participants, each with varying degrees of risk and profit participation. Such projects must satisfy government and public interests while permitting a satisfactory return to the private investors, constructors and operators. governments should be prepared to share risks with the private sector and should accept that very few privately financed toll road projects are likely to be financially viable without some form of financial contribution from the public sector as possible. The primary allocation of risk is determined through the concession agreement and the construction, operation, finance and revenue packages.

Government Financial Support

Governments should seek to minimise the need for public financial support for toll road concessions in order to maximise the benefits of the PPI projects relative to costs. In some cases government risk assumption and financial support may be necessary to support a project which would otherwise be unable to close financing because of weak projection economics or unfavourable country and concession environment. Government financial support may be appropriate, however, if it helps mobilise large amounts of private capital.

The Estimation of Traffic and Revenue Forecasts

Traffic and revenue forecasts are the underpinnings of the project and dictate project viability and the level of government support that will be required. These are, perhaps, the greatest risks faced by toll road projects. They are defined as risks associated with insufficient traffic levels and toll rates too low to generate expected revenues. The treatment of traffic and revenue risk ranges from full private sector assumption of the risk to government-provided minimum traffic and revenue guarantees. The risk mitigation measures are as follows:

- the independently verified ridership study;
- cash support in the form of guarantee of x per cent of traffic revenue;
- toll adjustment mechanism.

The Rapid Appraisal Method (RAM)

The RAM (Merna and Adams, 1994. pp. 95–8) has been designed to operate in conjunction with a structured concession agreement to perform a deterministic risk analysis for any type of BOOT (build–own–operate–transfer). The RAM is used to appraise the commercial viability of BOOT projects quickly. It is based on the project classification, the project packages and deterministic effects of risks. The RAM has obvious limitations in that it is deterministic and relies heavily on the user's judgement and ability to assess and cost the likely effects of risk.

Project Classification

Project classification is the basis for identifying project data requirements, the number of secondary contracts and organisations to be involved and many of the elemental and global risks to be considered in the appraisal.

Preliminary Estimate

The next stage of the RAM is to determine the profitability of the project based on the estimated costs of construction, operation and maintenance, finance and revenue generated. This is calculated in the preliminary estimate, the purpose of which is to establish an initial estimate of profitability, as follows:

1 Profit $p1 = R - (C + O + F)$

(where $p1$ is the profit, R is the revenue, C is the cost of construction, O is the cost of operation and maintenance, and F is the cost of finance).
   The estimated cost/profit ratio is expressed as

2 $r1 = (C + O + F)/ R - (C + O + F)$

The profit and cost/profit ratio are the first guides to commercial viability of the concession project. If the estimated profit and cost/profit ratio of the project estimate are acceptable to a private sector the next stage of the appraisal is to determine the component estimate.

Components Estimate

The major components of each of the packages can be expressed as:

- construction: $c1, c2, ...cn$;
- operational: $o1, o2, ...on$;
- financial: $f1, f2, ...fn$;
- revenue: $r1, r2, ...rn$.

(where $c1, cn$; $o1, on$; $f1, fn$ and $r1, rn$ are the components of the construction, operation, financial and revenue packages respectively).
   Initially the major components of the construction package are identified and allocated costs. Next, the operational components necessary to meet the

requirements determined under the construction package are identified and allocated costs.

The financial components required meeting the cost of the construction and operational packages are then determined.

Finally a revenue estimate is prepared based on the components of the revenue package.

The next stage is to re-determine the project cash flows and profitability, p2, as follows:

3   $p2 = R - (C + O + F)$

and the cost/profit ratio r2 is:

4   $r2 = (C + O + F)/R - (C + O + F)$

At this stage of the appraisal only the components costs and duration are considered. The effects of risk on the commercial viability of the project are only considered if the component estimate is acceptable. A project cash flow is prepared based on the cost of each component and timing over the concession period from which the IRR (internal rate of return), NPV (net present value) and pay-back period are calculated. The RAM calculates the pay-back period at which a positive cash flow is achieved. However, financial costs associated with interest payments will often be extended to the end of the concession period, as for the costs of operation.

The private sector determines a minimum acceptable rate of return (MARR) for identified project. The MARR for private companies is usually higher than that for public agencies, as private companies usually pay a higher rate of interest than the public sector does.

If the IRR of the component estimate is greater than the MARR then the effect of elemental and global risks are considered on the basis of the component estimate. If the IRR is equal to or slightly less than the MARR then the effects of inflation and the provision of the concession agreement on the overall project may be considered. If, for example, a project had a short construction period and toll levels were to be set by the private sector with freedom to increase toll levels then a further appraisal could be carried out. If the allocation of risks gives little or no scope for increase in revenue then no further appraisal would be sanctioned and the project abandoned.

Other Risk Mitigation Measures

- Identify implications of risk and who can best manage if needed, so that risk should rest with those most capable of handling it.
- Promote public hearings in the planning process and dealing with environmental issues.
- Consider a number of significant insurance policies in the context of concessions (e.g. contractor's all risks third party liability, employer's liability, workers' compensation, material damage, etc.).
- Develop the PPI projects, which requires the government to develop a policy framework for investment in toll road projects, clear criteria for selecting concessions, efficient and effective procedures for requesting bids and awarding contracts.
- The government must be committed to a particular project and certain of its economic viability before it asks the private sector to spend time and money preparing bids.
- To extract the maximum possible economic benefits from a privately-financed toll road project scheme, a balance needs to be struck between maximising the transfer of risk to the private sector, and thus the incentive to manage that risk effectively, and encouraging a sufficiently attractive risk to reward ratio to encourage private sector involvement and so on.

Government Policy for PPI Projects

If governments wish to encourage and facilitate PPI projects, they will have to provide an appropriate balance between risk and reward.

Developing countries must recognise that a higher level of risk will require a higher prospect of reward. If this is not forthcoming then governments must offer more financial support or accept a lesser degree of risk transfer than in countries which have developed financial systems and strong credit rating.

Governments can attract private participation in toll road infrastructure in two ways. They can offer financial support to investors – in the form of grants, cheap loans or guarantees – in order to compensate them for low tariffs, unstable macro-economic conditions, poor performance and other problems. Or they can address a policy to solve problems that underlie investors' concerns, raising prices to cost-covering levels, ensuring macro-economics stability and establishing a sound regulatory framework.

Government policies which facilitate PPI projects are as follows:

- changes in policy directions from a government-centred or government-financed scheme to a private sector oriented or private sector initiated scheme;
- promotion of creativity and efficiency from the private sector throughout the stages of planning, design, construction and operation of PPI projects;
- introduction of preferential treatment system for the private sector;
- provision of incentives for the reduction of construction period or project costs;
- restructuring domestic systems to meet international standards;
- mitigation of investment risks by providing a competitive rate of return, minimum operation revenue guarantee, foreign exchange risk reduction, etc.;
- evaluation of the project proposal and the designation of the concessionaire conforming to international standards;
- fair and equal treatment of domestic and foreign investors.

Conclusion

We have discussed the fact that financial reality can never be ignored for PPI projects through the two countries' cases. The feasibility study shows that a given percentage is needed for the given projects to attract public participation, even if politicians can push a project without any kind of government support. Otherwise, the final result can be a non-affordable toll level, while the public opinion can be much more adverse and financial consequences on the government can be much more severe.

At the time of writing, the Hungarian government seems to be avoiding PPI projects due to their high toll charges and political reasons. From January 2000, Hungary adopted the vignette system (considerably cheaper price), which is the licensing of motorway access on all toll motorways (M1, M3) except the M5. Under this system, vignette buyers can use motorways without limitation by attaching a valid sticker on their cars. The government's conception is to introduce this vignette system for all Hungarian motorways, including the M5.

In general, it is true that PPI projects need a much longer construction time and high toll charges compared with publicly funded roads. In spite of these weak points, interest in PPI toll projects is particularly strong because governments require alternative methods of financing their extraordinary transport needs. By using the PPI method, governments can use private capital and better management skills through the sharing of risks and responsibilities between governments and the private sectos. So mature economies, developing countries, and emerging markets all seek, each in their own way, methods to achieve common goals – economic growth, political independence, and more opportunities and higher standards of living for their people. To these ends, more and more countries are looking to the private sector to assist them in constructing the necessary infrastructure. Taking into consideration the financial difficulties of these regions, PPI is an essential element in Eastern Europe and Asian countries' highway development based on the public/private partnership principal.

Notes

1   Most countries have no toll roads. Where they do exist, the tolled network typically comprises less than 5 per cent of the road network. Tolls for passenger cars average around $0.03 to $0.08 per kilometre (the World Bank Internet Home Page: <http://www.worldbank.org>).

2   The toll rate is different from the size and type of vehicle. The toll is charged at 2,310 Hungarian Forint (about US$8) per ride in the case of car.

3   The toll rate also is different from the size and type of vehicle. The toll is charged at 6,100 Korean won (about US$5) per ride in the case of car.

References

András, T. (2000), 'Case Study on M1/M15', unpublished note, Budapest.

Hook, W. (1999), *The Political Economy of Post-transition Transportation policy in Hungary*, Transport Policy, Elsevier Science Ltd, Amsterdam.

Katalin, T. (2001), 'Basic Investigation Methods for Efficient Allocation of Sources in the Development, Operation and Maintenance of Transport Infrastructure', *Scientific Review of Communications*, Budapest.

Katalin, T. (2000), 'Euro Compatible Transport Infrastructure – Requirements and Possibilities. Hungary at the Millennium – Transport, Communication, Informatics', *Strategic Research in Hungarian Scientific Academy*, Issue: *Transport Systems and Infrastructures*, Budapest.

Katalin, T. (2000), 'Current Transport Technology Development in European Union Countries. Studies – Transport', *Strategic Research in Hungarian Scientific Academy*, Issue: *Transport Policy, Rail Development, Informatics*, Budapest.

László, H. (2000), 'Case Study on M5', unpublished note, Budapest.

Merna, A. and Adams, C. (1994), *Projects Procured by Privately Financed Concession Contracts*, UMIST, Manchester.

Ministry of Construction and Transportation (Korea) (2000), *Private Participation in Infrastructure Projects*, Seoul.

Perez, B. (1999), *Private Finance and the Expansion of the European Motorway Network*, The Diebold Institute for Public Policy Studies.

United Nations (1998), *Public/Private Partnership (A New Concept for Infrastructure Development)*, New York and Geneva.

World Road Association (2000), *Guide for New Methods of Financing and Public/Private Partnership*, Kuala Lumpur.

Chapter 16

Considering Multi-modal Capacity in the Assessment of Road Design

Markus Mailer

Introduction

The Austrian Standard for Assessment of Road Design (RVS 3.7, 1994) is in a process of revision at the moment. In Austria like in many other countries the current traffic situation is characterised by growing traffic on the roads, increasing financial problems of rail and stringent infrastructure budgets. The relevant department of the Austrian Federal Ministry for Transport has been well aware that the currently only road oriented standards are not a useful tool to solve these problems. It is perceived that their approach to adjust road capacity to traffic demand might even be a part of the problem.

The repercussions on traffic growth inherent in the level of service concept introduced in the American Highway Capacity Manual (HCM, 1985, 1997) has been described by Knoflacher (1994). The presented relationship explains findings like that stated in a recent analysis of a congestion study (SFPP, 2001), 'to respond to congestion by trying to add more space to the road system … has proven to be an inefficient strategy. … Even though road building has been outpacing population growth in metro areas congestion has grown worse in most places'. The reinforcing relationship and feedback loop between the level of service approach and traffic volume works like shown in Figure 16.1.

On the left side the figure shows the structure of the HCM and similar standards. Growing traffic volume is reducing travel quality. According to the standards, authorities take measures to increase road capacity when traffic quality falls under a target quality. Higher capacity is raising quality over the desired level again. On the right side of the figure the effects of this procedure on land-use and road traffic are illustrated. Travel quality is usually related to travel speed. Travel speed is related to land-use. In the last decades high travel speeds have caused urban sprawl, economic concentration and other developments resulting in longer travel distances. More road traffic is one of the consequences. Finally this will increase traffic volume on the road section

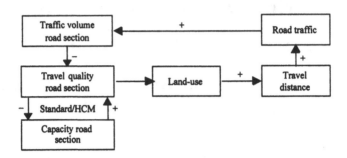

Figure 16.1　Reinforcing feedback loop between the level of service approach and traffic volume

which was the starting point of the loop. This relation makes clear why the usual response of road administrations to traffic growth, which is intended to increase road capacity, inevitably generates even more traffic. Due to the time lag in the relations between infrastructure improvement, land-use change and traffic growth, many still do not recognise this vicious circle. But both the growth of road infrastructure and traffic caused by this reinforcing loop have reached several limits already, not least a financing limit for infrastructure construction, operation and maintenance.

Thus the experts at the Ministry of Transport decided that the revision of the Austrian standard evaluating road design should result not just in an update based on recent findings concerning the influence of road design parameters on traffic flow and travel speed. Over and above this, it has to be extended to a multi-modal transport approach considering the capacity and quality of different modes on the road as well as in the corridor. This should result in a more efficient and cost effective use of infrastructure on the one hand. On the other hand, the new procedure should give a clear structure and transparency to the planning process and so provide a basis for the coordination of transport and land-use planning. In this regard the new standard should also be a tool that supports strategic decisions making on a higher level.

In a tendering procedure TUW-IVV[1] was commissioned by the Federal Ministry to do the research that should form the basis for the new standard. The research (Knoflacher, Mailer et al., 2001) was supervised by an expert committee of the Austrian Association Road and Transport which is preparing and publishing the Austrian road standards.

The Old Standard

The old Austrian standard for assessing road design is based on a typical demand-oriented HCM-like approach. It checks if a given traffic volume that was counted or predicted can pass an existing or projected road section under predetermined conditions. The average travel speed of the private car, which is called operational speed, was chosen as primary measure of traffic quality. Operational speed is dependent on road design. The standard procedure checks if operational speed exceeds a target speed that is dependent on road function (Figure 16.2).

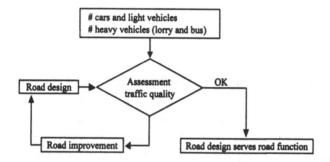

Figure 16.2   Concept of the old Austrian standard for assessment of road design

For road sections in non built-up areas the calculation of average car travel speed is based on a complicated traffic volume–travel speed algorithm. This algorithm considers design parameters such as number of lanes and their width, gradient as well as the load factor, i.e., the ratio of design volume to capacity. The used capacity value, however, does not reflect capacity in the true sense. For computing this value for a rural road segment an algorithm similar to that defined in the HCM for computing (service) flow rates is used. In the Austrian standard the parameters determining the capacity value are design parameters like number of lanes and their width, curvature, overtaking sight distances as well as the influence of heavy vehicles depending on their proportion in the traffic stream and the gradient. Due to the definition of the corresponding adjustment factors the capacity value includes the influence of these parameters on travel speed already.

For road segments through urban areas the calculation of average travel speed takes the prevailing speed limit as starting point. The distance to buildings or other lateral restrictions, the number of junctions, access roads and driveways as well as local restrictions of road width are considered.

For longer road sections with changing design, operational speed is calculated for each section and a weighted average has to be worked out including the lower speeds in urban areas and delay at traffic lights. Average operational speed has to be higher than a minimum speed fixed for defined road categories. If operational speed is too low, road design has to be changed (Figure 16.2). According to the parameters used this means that more or wider lanes and/or modifications of alignment and/or gradients are necessary. Since longer sections through urban areas reduce average operational speed, bypass roads were also a frequent measure to meet the speed target.

Changing the Target

In the process of reconsidering the old method the usefulness of defining travel speed as primary measure of traffic quality was also questioned. Problems related to road design in hilly or mountainous regions resulting in unreasonable alignment and expensive road construction had already been addressed, e.g., in Germany (Brilon et al., 1997). Practitioners also reported problems in justifying large-scale road reconstruction with little speed difference. Additionally, in the light of research proving constancy of travel time and travel budget (for example, Schafer, 1998) it has to be questioned whether travel speed is a suitable indicator if a road serves its function properly or not. Thus, in agreement with the supervising committee of the Austrian Association of Road and Transport, TUW-IVV decided to introduce the load factor as a new primary measure, defined as the ratio of design volume to the capacity of the road the load factor reflects the utilisation of capacity. Higher road categories should be associated with lower load factors, i.e., with traffic flows further away from the unstable traffic conditions at capacity. Of course, due to the fundamental relationship between load factor, traffic density and travel speed this still implies higher travel speeds for roads of higher priority.

Nevertheless, the new approach in combination with current research knowledge changed and reduced the factors influencing the assessment. For instance, within the usual range lane width is not an influencing capacity. Therefore this parameter is no longer relevant for the assessment. Obviously, the influence of slower heavy goods vehicles is not the same on capacity as on car travel speed. Similar considerations apply for curvature, speed limits and lower speeds in urban areas. Since roads rarely operate at volumes approaching capacity, high volume traffic data is difficult to obtain. Thus theoretical considerations had to be used to develop the new methodology. As a result the factors influencing capacity of rural roads could be reduced to the effect of heavy vehicles depending on their proportion in the stream and the gradient of the road. Thus the change of the target from speed to load factor not only simplifies the procedure, and so increases its user-friendliness, but might also help to reduce road construction costs. Local bottlenecks like traffic lights can be easily included by calculating their capacity and load factor. Longer road sections are still split into segments with uniform traffic and road conditions. But the reduction of the influencing factors results in fewer segments. All segments, all bottlenecks and all junctions have to be able to accommodate the design volume considering its actual traffic composition. For the assessment of the total road section a weighted average load factor is worked out, based on the lengths of the segments and the sections influenced by junctions or bottlenecks.

In the new standard the definition of design hour volume (DHV) is also reconsidered. This definition has always been a compromise between providing an adequate level of service for almost every hour of the year and economic efficiency. In the old standard DHV was derived from ranked hourly volume distribution and defined as the volume exceeded only in 30 hours of the year. This approach means that the capacity of the expensive road infrastructure is underused for more than 8,700 hours a year. Some countries – France, for instance – see road design in the light of economic capacity. Most OECD countries have been using lower values like the 200th highest hourly volume of the year, which has shown no negative effects on operation in field observations (OECD, 1985). Additionally, designing for the 30th highest volume of a year value is not suitable for multi-modal consideration, since this value cannot be compared to the values determining the operation of public transport. Usually, public transport operation is designed for the repetitive weekday peak. To create comparable conditions the revised standard defines DHV on the road with the average volume of the working day peak as well. If sufficient data is not available, design volume can be estimated as the volume which is exceeded in only 200 hours a year. If this value is not known either, design volume can be approximated to 10 per cent of average annual daily traffic (AADT). Considering the nature of traffic in the highest hours this default value can be adjusted. For international transit routes characterised by steady flows the 200th highest traffic volume might be as low as 5 per cent of AADT. This new definition of design volume is in line with the international trend of increasing the economic efficiency of road design.

Extending the Single-modal Approach

Changing the primary measure of traffic quality and definition of DHV already provides a basis for more efficient road design. But these changes still refer to a single-mode method like that of the old Austrian standard and other road standards for the assessment of road design that use vehicle units and personal car units respectively to describe traffic volume. Aiming at an approach that also includes other modes, it is obvious that volumes based on vehicles are not suitable. Vehicle units can neither be compared nor directly transformed to other modes. This means that the first step towards a multi-modal approach is to get back to the real measure of transport demand, hence looking at people and goods that are to be transported.

Recalling the transport purpose of traffic adds another dimension to capacity assessment. Capacity is defined in terms of the maximum combination of persons and goods that can be accommodated by a given traffic facility under prevailing conditions. This approach not only considers the capacity of the infrastructure itself, but also capacity of the means of transport. For road transport car occupancy and the loading of goods vehicles gain importance. So this approach shows different ways in which road capacity can be improved. It allows the consideration of traffic management measures. Measures which are able to improve car occupancy or loading of goods vehicles can increase transport capacity in terms of persons and goods without increase of road capacity in terms of vehicles. So this first step of extending the vehicle-based single-mode approach already shows alternative ways of meeting a load factor benchmark. If capacity and design volumes are defined in passenger and goods units, useful and effective traffic management measures can increase road capacity and so decrease the load factor for a given design volume. This adds various scenarios to road improvement measures and so provides different options to react on failing the road traffic quality benchmark. These scenarios comprise management of goods and management of private car (PC) traffic (Figure 16.3).

Since roads always have to serve both passenger and goods transport both scenarios are closely related. When maximum flow of vehicles is limited to meet traffic quality targets, measures related to both scenarios can help to accommodate the combined transport demand. Improvements of car occupancy, for instance, can result in free transport capacity for more passengers or goods or both. If, on the other hand, management measures succeed in a more efficient goods transport, hence in reducing the number of goods vehicles needed to transport the given quantity of goods, additional capacity for passenger transport is gained. In both cases the same number of vehicles can transport more passengers and/or goods.

The different scenarios are, of course, not equal in respect of effectiveness, efficiency and chance of realisation. Therefore it is necessary for all scenarios to identify those measures or sets of measures which can make the scenario meet the benchmark. Finally a feasibility and a cost-benefit analysis should be applied to find out which of the measures could be realised and which of them would be most effective in bringing traffic quality to the desired level. Many traffic management measures are related to origin and destination areas of trips rather than to the road section that is traversed (e.g., parking policy). For these measures cross-sectional demand analysis is inappropriate. The procedure has to be extended to an origin-destination based method. This requires a much wider database and a suitable transport model. The model which is calibrated on the present situation has to be able to compute the effects of different measures of infrastructure extension and traffic management on the flows of passengers and goods on the road. This shows that even in a single-modal procedure, the change of the transport paradigm alone results in an extended demand for data, tools and expertise. Or, to put it another way, more effort has to be put in the preceding planning process to find the most efficient way to achieve the target traffic quality.

This applies even more when it comes to multi-modality. It is generally assumed that the consideration and coordination of different modes will result in a more efficient transport system. But, of course, there is a trade-off between cost savings in realisation and planning effort. The much more complex multi-modal consideration implies a need for more local data and information about the nature of peak traffic to make informed judgements.

In the assessment of roads the first level of multi-modality refers to public transport on the road, hence to public bus services. To identify the potential of modal shift it is necessary to know the origins and destinations of, and time constraints on, the people using a specific road section. Measures that succeed in making people change from car to public transport increase bus occupancy and hence vehicle utilisation. So once again the same number of vehicles can transport more passengers and goods. This adds a further scenario for making road transport more efficient, i.e., the management of public transport (PT) (Figure 16.3). This scenario might be of special interest to those countries in Southeastern Europe which face growing motorisation and a mode shift from bus to private car. This trend makes road transport less efficient and might cause problems in the near future. Scarce resources might be saved if appropriate remedies are taken at an early stage.

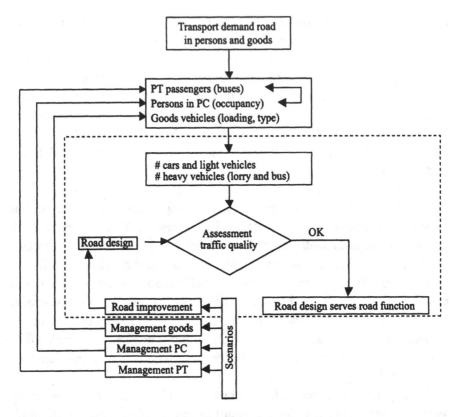

Figure 16.3   Extended concept considering transport demand and capacity in passengers and goods allowing for different management scenarios of a more efficient road transport

Extending the Assessment from the Road to the Corridor

To complete the multi-modal approach the concept is, finally, extended to the corridor. The corridor comprises the road section that has to be assessed as well as all parallel means of transport. The whole transport demand in passengers and goods using this corridor is taken into account. Corridor capacity results from the capacity of all means of transport. But, since the new standard is still a road standard, the assessment focuses on road capacity utilisation. The transport contribution of the other means of transport is considered as reduction of the transport demand on the road. The share of the transport demand which is accommodated by other means of transport can be deducted from the corridor demand before the actual assessment of the road. In this concept the road has to accommodate the demand which is not accommodated by other means of transport. Consequently, when the road fails its traffic quality target, this can be interpreted as insufficient performance of the alternative means of transport. This approach results in another scenario for making transport more efficient, i.e., management of modal split (MS) in the corridor (Figure 16.4).

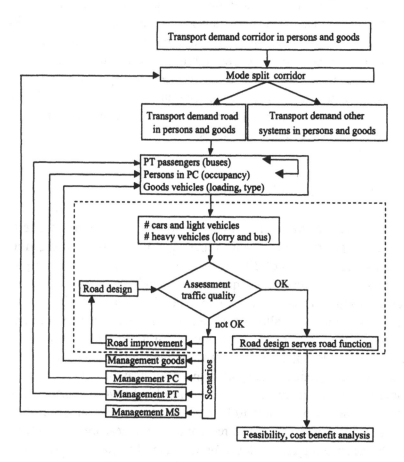

Figure 16.4   Further extended concept including all means of transport in a corridor and different modal split scenarios

For this multi-modal approach a suitable transport modal is needed to identify the potentials of mode shift in the corridor and to estimate effects of appropriate measures. It has to be based on data about origin, destination and time constraints of passengers and goods allowing the calibration for the present situation. The model has to be able to compute the effects of various management measures on mode choice. Therefore it has to consider all parameters describing transport quality of the different modes, including mode availability and accessibility, travel time, costs, reliability and comfort. Based on the model, different scenarios to meet the traffic quality target on the road can be worked out. Due to the multi-modal concept suitable measures might be related to other means of transport as well. Measures improving capacity utilisation of rail, for instance, would result in better capacity utilisation of the corridor. Investments in rail capacity might also be suitable to reduce transport demand on the road, hence to ensure adequate quality of road traffic. Considering the heterogeneity of suitable measures related to the different scenarios, feasibility and cost-benefit analyses have to be applied to identify most effective and most efficient measures.

Developing the New Approach Further

The extended concept of road assessment allows multi-modal capacity utilisation and capacity management to meet quality targets of road traffic. It supports the identification of different scenarios indicating how transport demand can be accommodated adequately. It extends the range of suitable measures to respond to the unsatisfactory quality of road traffic. Apart from road improvements, it allows for management measures controlling a given demand rather than merely accommodating it. Still, the procedure is demand oriented. It does not question transport demand, nor does it look for measures to usefully reduce transport demand by changing travel pattern. This would require the inclusion of the relationship between transport and land-use. The model would have to be extended to an integrated transport and land-use model considering the effects of land-use measures on transport demand. On the spatial planning level the function of a road or corridor can be reconsidered as well. The research undertaken by TUW-IVV pinpointed the need for this further development in the assessment of roads and other transport infrastructure.

Conclusion

In a situation of growing transport demand and limited funds for infrastructure improvement, the assessment of road design has to be reconsidered. In Austria the revision of the relevant standard is using two tracks to make transport more efficient in order to ensure adequate quality of road traffic. On the one hand, making capacity utilisation the primary measure of service quality already results in more efficient road design. Compared to a speed-oriented approach an assessment based on capacity reduces the influence of design parameters such as horizontal alignment and grade, as well as of the traffic mix (proportion of heavy vehicles).

The change of transport paradigm, however, adds a further dimension to the assessment of capacity utilisation. In order to allow for multi-modal considerations, transport demand has to be measured in persons and goods rather than in vehicle units. For the assessment of roads this combines the utilisation of traffic capacity in terms of vehicles with the utilisation of vehicle capacity in terms of occupancy and loading. So the approach allows the consideration of the characteristics and potential of public transport on the road and of alternative means of transport in the corridor. Different scenarios can be worked out to accommodate transport demand more efficiently. To identify the measures related to these scenarios a suitable transport model is needed. It has to be calibrated with origin- and destination-based travel data. The evaluation and comparison of scenarios requires feasibility and cost-benefit analyses of the related measures.

Due to the complex mechanism determining modal split and mode shift potentials the consideration of multi-modal capacity in road assessment results in a substantially increased effort demanding adequate models and expertise. However, the application of a multi-modal approach makes sense for the assessment of large infrastructure projects in particular. The outlined concept provides a clear and transparent structure for the planning process. It pinpoints unutilised capacity and allows effective identification and efficient measures to ensure adequate service quality. Thus it provides a useful decision-making tool for the allocation of limited funds, especially for governments who are confronted with claims for investments in different modes.

Note

1   Technische Universität, Institute für Verfkehrsplanung und Verkehrstechnik (Institute for Transport Planning and Traffic Engineering, Vienna).

References

Brilon, W., Grossmann, M., Blanke, H. (1994), 'Verfahren für die Berechnung der Leistungsfähigkeit und Qualität des Verkehrsablaufes auf Straßen', *Forschung Straßenbau und Straßenverkehrstechnik*, 669.

HCM (Highway Capacity Manual) (1985, 1994), Transportation Research Board, Special Report 209, 3rd edn, updated 1994.

Knoflacher, H. (1994), 'Do We Use the "Level of Service" Concept in the Right Way?', 2nd International Symposium on Highway Capacity, Australia.

Knoflacher, H., Mailer, M., Schopf, J.M. and Fischer, P. (2001), 'Multimodale Dimensionierung von Straßen', report commissioned by the Austrian Federal Ministry of Transport.

OECD (1985), 'Leistungsfähigkeit von Hauptverkehrsstraßen' ('Capacity of Major Routes'), *Forschung Straßenbau und Straßenverkehrstechnik*, 445.

RVS 3.7 (1994), 'Überprüfung der Anlageverhältnisse von Straßen. Forschungsgesellschaft für Verkehrs- und Straßenwesen', Vienna.

Schafer, A. (1998), 'The Global Demand for Motorized Mobility', *Transportation Research Part A*, 32 (6), pp. 455–77

SFPP (Surface Transportation Policy Project) (2001), 'Easing the Burden, A Comparison Analysis of the Texas Transportation Institute's Congestion Study', <www.transact.org/Reports/tti2001/report.pdf>, 13 June.

Chapter 17

Sustainable Cities for Countries in Transition: Learning from Mistakes in Countries with High Motorisation

Hermann Knoflacher

Introduction

Cities in western Europe are confronted with increasing problems not only in the transport sector but also in the environmental field, increasing social problems, city sprawl and loss of economic power. Economic activities are leaving the cities and settling along motorways, offering free parking opportunities for their customers. On the other hand, public transport, in former times the backbone of the transport system, is losing its customers and becoming a highly subsidised system. The increasing congestion on all carriageways for cars, particularly on motorways, is a clear symptom of a failing in the market economy in this sector. All kinds of treatment to mitigate or solve the problems have failed so far. The traditional approach seems to be totally wrong. The system has not been understood. A new approach, taking into account on one hand, individual human behaviour – which has not previously been the focus of traditional transportation planning – and, on the other hand, system behaviour, which has never been considered (although early studies from the end of the nineteenth century (Lill, 1889) point in this direction), offers the basis for understanding these unexpected effects. Countries in Southeastern Europe now have opportunities to avoid the main mistakes in city planning and transport system development, if they apply the scientific and empirically-sound findings. Motorisation in these countries is still much closer to a sustainable level than in most of the cities of western Europe. Although Newman and Kenworthy (1989) have no data from these cities in their diagram, it can be taken as given that most cities of Southeastern Europe are still in the sustainable position, considering density and energy for transportation.

Their chances of chosing a more sustainable path towards the future are much better than for most cities in western Europe. This chapter will describe the new approach towards a sustainable, socially balanced and economically powerful city development.

Key Words and What They Mean

Sustainability and quality of life have become key words in many research areas in the 4th and 5th Framework Research Programmes of the European Union. They are also of increasing importance for practitioners and politicians in transport planning and city development. But what do these expressions mean in reality?

- *Sustainability* requires stability, resilience and flexibility over time.
- *Quality of life* has several definitions: quality for people today, or in the future? Quality of human life or life in general, including humans?

  Quality is not something which appears suddenly. Sometimes it needs a very long time to develop – and the development of quality is a process. Quality is the final result of an optimisation process, which always needs much more time than the life span of the elements of a city's system.
- *Time* has now been mentioned twice. What is a 'long time' in the context of our subject? The minimum length of time to reach some stability in a process is the lifetime of the core element – here the human being. The assurance of stability not only in our lifetime but in the future is necessary to test the resilience of the development against changing and fluctuating conditions in the environment, and in our case also in the economy. If a process is developing faster than this critical time, the risk of instability and collapse increases progressively.

Complexity of Technical Systems

A problem in modern societies is the complexity of the system they have built with their technical means. This system has a behaviour which is no longer directly visible and not easy to understand from the analytical point of view.

The system is an open system, depending on material and energy flows. The behaviour of the system's elements is not at all linear and, finally, it changes over time – it is dynamic. The system behaves like a natural system in its complexity.

Why do Natural Systems Work Perfectly?

Natural systems work perfectly because they have had time to develop perfection – more than four billion years. The optimisation process was so long that we are finally able to think in terms like 'sustainability' and 'quality' since we are able to recognise these attributes in our environment.

Contrary to this development, modern techniques and the needs of economies are developing much faster and do not have the time to wait until the system they build has developed to a sustainable stage. There is no time to wait for enough experience to determine whether the solution chosen is the best one for the future; often it is good enough if it is good enough for today, or even for yesterday. Acceleration is the credo of this period. The probability of developing quality or sustainability in such an accelerated process is very low, close to zero. As sustainability is always the result of high quality, structures developed in such a process of haste are not at all sustainable. So quality is an important indicator for sustainability.

Mobility and Time Constraints

Is the mobility of today a contribution to life, which is dependent on life-supporting systems – physical, social and cultural – or do some types of mobility endanger quality of life and, therefore, sustainability?

If we look at different parts of the modes of transport we use today, we have enough experience with walking. The whole of human development reflects the development of the biped, walking in upright position on the earth for more than 2 million years. For about 5,000 to 10,000 years boats and horses have been in use, and with the development of these types of transport, settlements appeared. Two hundred years of bicycling, about 150 years of railways, 100 years of cars, about 50 years of aviation and some years of telecommunication characterise the rapid development of the transport field.

Hypothesis: 100 Years of Car Use are not Long Enough to Produce Sustainable Quality of Life and Sustainable City Structures

The quality of a complex system like a city, can only be described by well-defined indicators. Indicators of the modern technical transport system are far from acceptable areas. Examples are:

- 1 hour's daily use of cars by car drivers produces 24 hours of noise and air pollution;
- 1 hour of comfort for 24 hours of pain does not provide the right cost/benefit ratio;
- space and energy indicators point in the same direction
- the ecological footprint of this kind of mobility is too big.

But: Mobility is a Prerequisite for Life and Quality of Life

Is this true for each kind of mobility? Physical mobility was expensive before the invention of transport modes driven by external energy sources. Mental mobility is always expensive, so people try to avoid this effort. Individual quality can be defined as an effortless, enjoyable, interesting, safe and hopeful kind of life, but this cannot be sustainable. There is a trade-off between mental and physical energy (Figure 17.1).

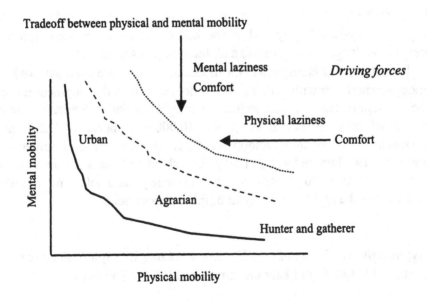

Figure 17.1   Trade-off between physical and mental mobility

Problem of Open Systems

To prevent degradation, we have to pay the price for negentropy (i.e., to keep order and prevent entropy). Technical means in the transport system have replaced human or animal energy by fossil or electric energy. The user has the impression that he or she can save energy and can move effortlessly. But in the system the demand for energy has greatly increased. For the same purpose, 100 times more energy is used and this has enormous effects not only for the transport system itself but also for many other structures. New mobility, based on external and cheap energy which replaces body energy, has created changes in:

- structures of settlements and economy (nearly all kinds);
- societies;
- cultures;
- values;
- the definition of 'quality', which has been reduced during the last 50 years to:
  - quality of traffic flow; and
  - quality of transport.

Even the lowest indicator of sustainability – keeping humanity alive – has become obsolete. Ensuring a safe and socially-encouraging environment for the next generation has become less important compared to convenient parking and being able to use the car. The analysis shows that mechanical kinds of mobility show all the signs of violating even the basic elements of sustainability:

- they have too large an ecological footprint;
- they promote cultural degradation – not just change;
- they lead to the degradation of system supporting values; and
- they lead to the degradation of settlements and urban structures, etc.

Effects of This Kind of Technical Progress

The availability of cheap external energy for transport gave and continues to give almost unlimited opportunities to save precious mental energy for nearly everybody (Figure 17.1). City planners can forget all their duties to create a liveable city.

The 'Athens Charter' was used in a way which separated and isolated city functions from each others. The multi-functional city was divided into areas for housing, working, recreation and central functions, all in different places, producing a tremendous amount of mechanical traffic.

- poor city planning creates problems in the transport system:
- politicians are happy: they can 'solve' local problems by moving them away – out of their area of spatial responsibility. Economy can fulfil the basic theories: 'economies of scale' and the miracle of 'theory of comparative cost advantages' perfectly;
- car drivers could compensate for local deficits by using the car;
- transportation planning became a highly respected discipline;
- car and oil producing industries become a synonym for the national economies;
- car driver and transport building lobbies acquire more and more political power to transform the environment in order to fulfil the wishes (not needs) of their clients.

GDP becomes the leading indicator for all decisions and this system fits very well into this set of dogmas, such as:

- saving time by increasing travel speed;
- increasing mobility;
- freedom of modal choice, etc.

Modern transport science knows that all three assumptions are wrong. There is no time saving in the system, if speed is increasing. The time budget remains the same, only the distances increase. Mobility in terms 'trips per inhabitant and day' has not changed, although motorisation has increased. There is a change of modes but not number of trips. And finally there is no absolute freedom of choice for real people. External circumstances (built structures, lack of information, etc.) hinder people in this kind of freedom.

Why has This not been Well Understood in the Last 200 Years?

On one of the latest levels of evolution – the artificial environment – the car has become extremely powerful in one of the earliest and therefore most effective human achievements. Between 50 per cent and 85 per cent of body energy can be saved by using a car as compared with walking. At the same time the speed of effortless movement increased tenfold and more. Freedom from time and space was possible – or seemed possible – the ancient dream of mankind was nearly fulfilled. The user got all the benefits; negative effects did not touch the senses directly! The causes were covered under thick layers of scientific, social and cultural taboos. Instead of science, ideology and a mistaken picture of humanity became dominant. The creator of this wonder, humanity, could not be victims of their own invention. This belief still dominates people's thinking and worldview. If elements of prior importance are ignored, measures using elements of secondary effectiveness can not be successful. This seems to be the case in most of the so-called 'best practice' collections concerning urban and transport management. Most of them are arrangements in some way of factors and measures of different levels and different importance for the system.

Human Behaviour and its Background – the Key Element

If a society invests large amounts of money into structures and infrastructures which create behaviour with increasing adverse effects, as in the transport sector, and then tries to increase the costs of use of the system, this looks more than contradictory for an outside observer. First the resistance is reduced by physical measures by building extensive roads, then it is increased through financial measures by introducing road pricing when congestion occurs.

This is the situation in many cities today. On the other hand, public transport has to be subsidised, although its indicators conform to urban needs much better than cars' indicators do. Obviously we have not understood human behaviour well enough to handle it in the right way in the artificial environment we have built during the last two centuries.

If we do not understand human behaviour much better than in the past, we can dream about liveable cities of tomorrow, but can never reach this goal.

Myths and Dogmas instead of Scientific Rationalism

Humanity has about 8 million years' experience with pedestrians. This time might have been long enough to understand the effects of this kind of mobility on culture, civilisation, nature, economy and to develop a kind of responsible behaviour in the society. One of the results was the development of cities, not only as physical structures but also as economic entities and social bodies. However, during the last two centuries a dramatic change occurred, when we began to use external energy for the movement of people and goods – and information. The pace of the process was so fast that hardly any of the disciplines dealing with it had time to understand what had happened and what effects could be said to be the results of these changes in the traffic system. Everything that could contribute to more speed less personal effort and increased personal convenience was welcomed. A new era of city and transport planning arrived. New dimensions were necessary to fulfil the needs of the mechanical transport modes. The human dimension and the restrictions of the past seemed no longer valid. Roads were converted into carriageways and the local multi-functional structure of the city was no longer necessary.

Transport planners built models which calculated the time savings from high speeds in the new mechanical transport system.

This new phenomenon gave new opportunities for city development and transport infrastructure. Technical means rather than human needs became the dominatant factor in city planning and management. But new and unexpected effects occurred: accidents, noise, air pollution, city sprawl, the dominance of corporations, congestion, the death of small economic structures, financial problems for cities in increasing subsidies for public transport, etc.

The planners did not recognise that their planning was based on beliefs and not proof and on dogma instead of on sound theoretical and empirical grounds.

Three dogmas dominate:

- the growth of mobility;
- time saving by increasing speed;
- freedom of choice for the system user.

None of these dogmas exist in reality. There is no growth of mobility, in terms of 'number of trips'. There is only a shift of mobility from one mode to another. Nowhere on earth has time saving by increasing travel speed been observed. Increasing travel speed inevitably has two concurrent effects: city sprawl for housing and concentration of economic activities.

Finally, travel time for pedestrians, cyclists and car drivers remains the same. The difference in travel time produces only differences in travel distances. The calculation of benefits for investment into faster transport modes is based on assumptions which are rooted in individual experience but not on real system behaviour.

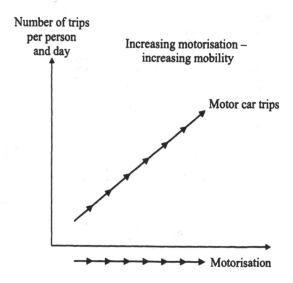

Figure 17.2   The myth of growing mobility

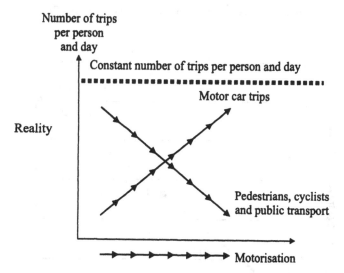

Figure 17.3   Shift between modes

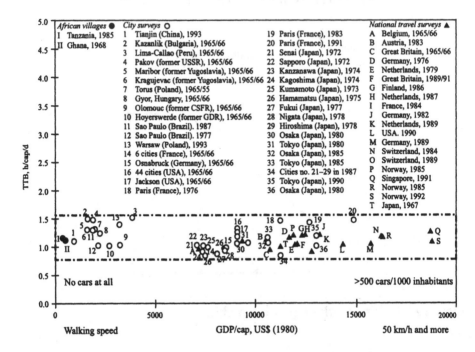

Figure 17.4   Constant travel time budgets (Schafer, 1998)

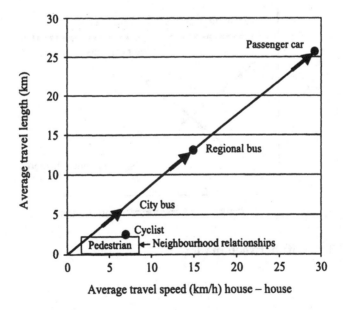

Figure 17.5   Distance and travel speed

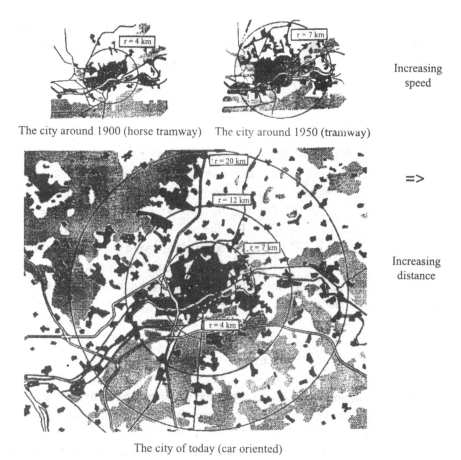

Increasing
speed

The city around 1900 (horse tramway)   The city around 1950 (tramway)

=>

Increasing
distance

The city of today (car oriented)

Figure 17.6   Urban sprawl (Wortmann, 1985)

The myth of 'freedom of choice' is based on a picture of a person who does not (yet) exist.

The effect of mechanical transport systems on human behaviour, human scale of values, human culture and social systems was not taken into consideration. It was forgotten that the city building elements are people. People are two-legged, walking in an upright position and not 'four-legged' like car drivers. Only pedestrians form a city on a human scale. Car drivers form a city on the scale of machines, which is not a human one. City planners and traffic engineers as well as economists and politicians see the problems in the traffic flows and try to mitigate or solve the problems in this field – but with no success at all. Cities become more and more unsustainable. They are wasting time, effort and taxpayers' money. They do not know that they treat symptoms and not causes (and there is no difference in ignorance between a small city or the European Union/Commission).

Parking: The Forgotten Key Element

Since parked cars do not move they have not attracted the irritation of the experts. Parking is regulated in a static way in guidelines for city planning and transport, but very seldom seen in its dynamic dimension. This can easily be explained with an example: large suburban shopping centres are a problem for the survival of inner city retailers. But the problem is not the centre itself, it is the parking supply it offers. If all parking places were cut off, the problem of shopping centres on the periphery of cities would disappear.

Parking is therefore the key element for the development of new and the revitalisation of existing cities.

Human Behaviour and Parking

It is necessary to understand what is happening when we can use the car. The car is the invention of our civilisation, one of the latest levels of the evolution of humanity. But are the effects of the car limited to this level?

If this were the case all the effects we have noted would not occur. The car would be controlled by society as well as by the individual. There would be no city sprawl, and society would not allow the killing of thousands of people in car accidents: the car would become extinct like animals in danger.

As was discovered more than 20 years ago (Knoflacher, 1981), the car influences human behaviour at our oldest and most powerful level evolution – body energy. It changes this basic and fundamental level and therefore all levels above, too. So it is easy to understand the dramatic change in human values in a car-oriented society. Such a society can never develop a sustainable city, nor keep it. Convenient mobility in a sitting position is much more important than all benefits of a liveable city, the responsibility for the future. Humanity is trapped in the car, becoming a four-legged creature, as we might have been 8 million years ago.

The question is: 'How can we escape from this trap?' The scientific and practical answer is: first it is necessary to know the construction of the trap and then to develop a procedure to help people to escape towards a sustainable

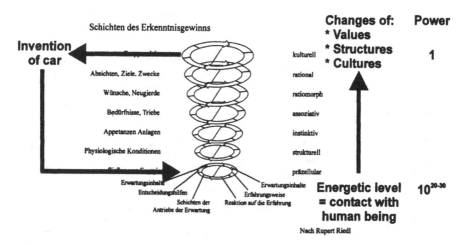

Nach Rupert Riedl

Figure 17.7 Evolution and the car

city. The construction of the trap was found in the human perception of the environment in the basic law from Weber-Fechner (1836, 1871) which was introduced in transportation planning in 1981 (Knoflacher, 1981):

Logarith and exponential functions are driving the system – inevitable

Sensation = ln(intensity of irritation)

$$S = \ln I$$

$$\text{or } I = e^{*S}$$

Sensation can have a '+' or '–' sign

Figure 17.8 shows the decrease of acceptance with increasing length of a walkway.

If cars are parked close to human activities as they are today, nobody who has the opportunity to use the car will choose public transport, if it is further away, as shown in Figure 17.8.

When someone has chosen the car, they are lost to the neighbourhood and the shops nearby, and is interested only in finding a convenient parking place at the end of the trip.

If we analyse real human behaviour in a car-free environment, people accept distances on the same level of acceptance which are more than 70 per cent longer.

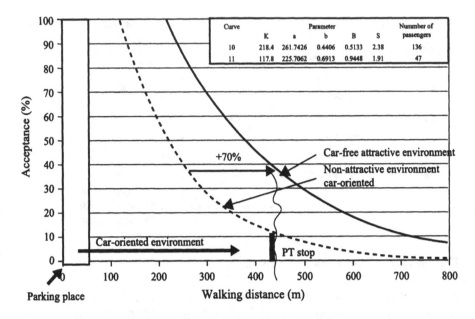

Figure 17.8   Acceptance of walking trips

The parking regulations which are in use today destroy each base for a sustainable city life.

If people have their car close to their activities, they need no activities close to their homes – city sprawl is inevitably the result.

The Solution

Now we know the 'construction of the trap'. If we want to escape we have to introduce a mechanism which gives us the choice of developing a sustainable city. The minimum condition is to remove the cars from human activities at least as far away as the public transport stop is located.

Provision for parking is no longer connected with housing or local activities. It is the key element to solving most of today's problems in our cities. Provision for parking has to be made by an independent organisation, which is also responsible for the functioning of the other parts of the transport system. Apartments, shops, workshops and other activities have to be totally disconnected from any parking provision. If someone wants to use and park a car they have to look for parking opportunities on the market, which has to be a market with the same opportunities for public transport. Nowhere, either

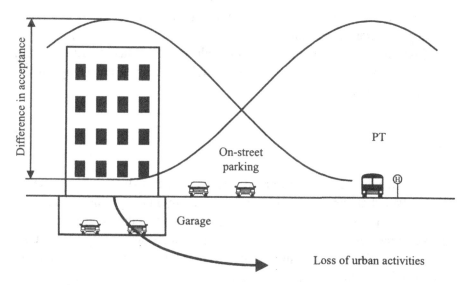

Figure 17.9   Today's parking regulations

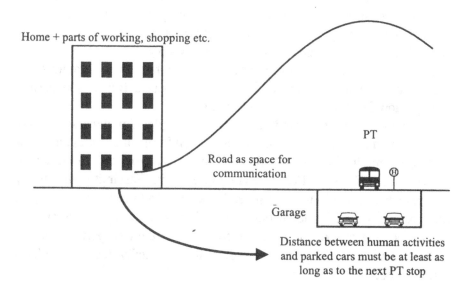

Figure 17.10   Key to the solution

in the city or in the village, should a parking place be more easily accessible than a public transport mode.

This is the minimum requirement for a balanced settlement structure, which is the basic precondition for a realistic (and not only virtual) way toward sustainable cities.

Effects

It is clear that this scientifically based result is not easy to realise. The mistakes are so basic, that they have to be repaired not only in the physical structures, but also in the financial structures, the guidelines and the principles of city management. There is no substitute for the precious body energy of man than body energy. The first step into this direction is the amount of mental energy as part of body energy to understand the mechanism of the trap, this means to understand human behaviour better, as a precondition to understand cities better.

The effect on settlements in highly motorised countries is well known today: loss of economic power in cities and communities is the outcome of this degradation of structures. First small structures disappeared, then middle-sized one and finally competition between the remaining large structures began, by increasing the speed of the system wherever possible.

As soon as people were able to use the car they had access to more speed and many advantages compared to slower system users. But with increasing motorisation the system changed the structures.

Europeans have followed American principles for decades. Today American planners are coming to Europe to learn from our old structures how to escape from the dilemma that this way of life has produced. In the annual Transportation Research Board meeting of 2001 in Washington in January these attempts were discussed at several meetings. But most of the examples are questionable, since the encouraging early results are not stable enough to convince the observers. We can recognise that the transport system (or the mode) is influencing the city structure fundamentally due to the effects described before.

The Way to the Solution

What is the Solution?

If the underlying, system dominating structure in which the basic system behaviour is rooted does not change, no sustainable solution is possible or effective. If the cause is the energetic structure, the measure must come into power at the same level if it should be effective. Any measure which does not touch the level of human body energy might or might not contribute, but it cannot solve the problem.

Existing Structure

Under given conditions car owners are not at all interested in using public transport to visit the local shops, etc. They are captured in cars and no longer connected to the community or city structures. Neighbourhood is losing its attractiveness and the city centres and traditional living districts are downgraded. The cause is not the traffic flow, rather it is the existing organisation of parking.

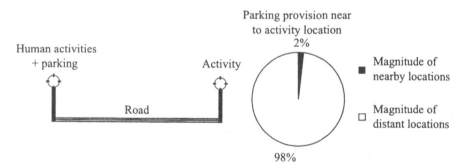

Figure 17.11 Effects of existing structure

The Basic Solution

Only if the parking organisation is changed can the system can be stabilised without loss of mobility and accessibility – for everyone. The distance to the parked car must be at least as long as the distance to the public transport stop. Cars must to be parked in garages or parking places at a distance no less than the distance from the public transport stop.

Then people have the chance of choice between modes and many activities will return the city. Market economy has now been introduced into the system of city and transport, which has been forgotten for the last 100 years in the west. Car traffic was excluded from the principles of market economy in many ways. The main element is parking organisation and not the traffic flow which was in the focus of many so-called traffic experts in engineering and economy. If this principle is understood, it becomes clear why all attempts to introduce road pricing will have no substantial effects.

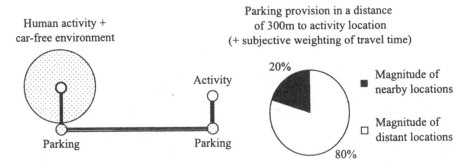

Real effects, taking into account human behaviour: 80% of the city is revitalised

Figure 17.12 Effects of parking organisation

Situation in Southeastern Europe

In most of the cities in Southeastern Europe the situation in term of motorisation is much better than in cities of European Union countries. There are excellent opportunities to prevent mistakes in the transportation system which have produced nearly insoluble problems in highly motorised countries. In European Union countries motorisation is between 400 and more than 500 cars per 1,000 inhabitants. In Croatia, Bulgaria and Hungary motorisation is between 200 and 250 cars per 1000 inhabitants, and in Rumania around 130 cars per 1,000 inhabitants. This give politicians and administrations the opportunity to develop city structures in a much more sustainable way than in the west. The preconditions are still much better, if the responsible people are clever and brave enough to apply the results of the system research which have shown the fundamental mistakes made during the last 50 years in western Europe, by copying American transportation planning principles from the 1930s. In the meantime, Americans are looking for better solutions in Europe. They will find much more in the countries in transition than in European Union countries. Cities in Southeastern Europe have the possibility of keeping the modal split in favour of public transport and developing sustainable economic and environmental structures, if they start immediately to bring order to the key element of city development – parking places. If they organise parking as described in this chapter they will overtake the development of western cities, if they also introduce market economy into the transportation system, especially into car traffic. This will prevent exponential growth of state deficits not only in the transport sector, but also in other sectors. If these countries fail to develop the transport system for their people and follow the western path of transport development they will soon be controlled by international corporations, whose power is dependent on cheap and fast transport systems. The burden of cost has to be carried by the people, while the corporations gain control over the cities and countries by means of road transport .

The structures described in this chapter have already been applied in several cities in the European Union, such as some parts of Vienna, in Eisenstadt and some others. Everywhere they have lived up to their promise. But they need excellent professional skill from the administration, system knowledge from the experts and brave and far-sighted responsible politicians.

Knowing the system effect is only one side of the coin. Applying the principles in the right way and realising the system in a proper way is the other: a much more difficult proposition.

The European Union will not support this sustainable development, since it will be some time before the Commission is be able to understand the basic principles of the transport system of today and tomorrow. In the European Union, transport systems are still treated with by applying the old and mistaken myths of time savings, mobility growth or time losses from congestion. As long as this pre-rational approach to the transport system is dominant, the wrong measures will be supported. The best example is the traditional approach towards traffic problems by treating the flowing traffic, which is only the symptom of the real cause – the wrong organisation between structures and transportation modes at the origins and destinations of all trips. This is an area where the countries of Southeastern Europe can demonstrate how future-oriented transport solutions should be organised.

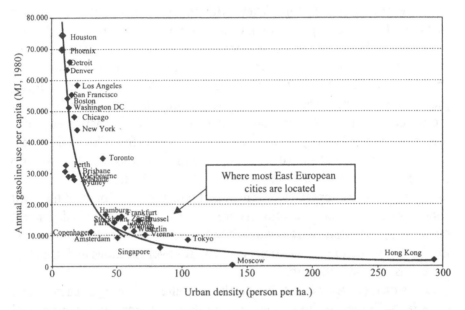

Figure 17.13 The existing situation in cities of Southeastern Europe gives great opportunities for sustainable development

Conclusion

In a dynamic system interrelations between the elements are the crucial point to keep it sustainable. The system depends on its elements and the elements depend on the system: humanity – society – manufactured structure – nature. Quality attracts people – and generates mobility. To be sustainable, the footprint has to be (at least) smaller than the bearing or carrying capacity of nature, man, society, culture, etc.

The Outcome: Sustainable Micromobility

There are many well documented examples where the efficiency of this principles have been quantified. Some of them are also included in European Union research projects.

Micromobility was the life-blood of all sustainable cities. If we can introduce this lifeblood into our cities by organising the transport system in this way, the cities will become again the centres of our economy, culture and society – and they will be sustainable, as they have been during the last 10,000 years.

Since the effects of an uncontrolled motorisation have not yet caused too much damage in most of the Southeastern European cities, they have the opportunity to reach the goals of sustainable development better than most of the cities in highly motorised countries. This will only be possible if the administrators and politicians are able to implement future oriented city and transport planning principles, as described in this chapter. If they follow the path of uncontrolled car motorisation, which offers individual opportunities, they will end in a disaster not only in the transport system, characterised by congestion on the roads, exponential growth of deficits or subsidies for public transport, city sprawl and loss of control for city development to international corporations. Hand-in-hand with this development they will also lose their identity and their cultural originality.

It is obvious that the principles described in this chapter are much more difficult to implement, since they take into account the whole system and its needs and not only the individual advantages of some persons or influential lobbies. They need system knowledge and excellent professional skill which are not yet available everywhere. But the results of the realised examples show all the benefits of this new approach, which is rooted in good European scientific principles and not in American superficial assumptions which have dominated the transport system during the last 50 years. Southeastern European cities have the opportunity to prevent fundamental mistakes which happened in western Europe by thoughtlessly copying primitive American principles of transport and city planning, which caused tremendous problems today in these historical and culturally highly developed structures.

References

Knoflacher, H. (1981), 'Räumliche Wirksamkeit von verkehrsorganisatorischen Maßnahmen', *Berichte zur Raumforschung und Raumplanung*, 25, Jahrgang, Heft 1.

Knoflacher, H. (1981), 'Human Energy Expenditure in Different Modes: Implications for Town Planning', *International Symposium on Surface Transportation System Performance*, US Department of Transportation.

Knoflacher, H. (1996), *Zur Harmonie von Stadt und Vekehr – Freiheit vom Zwang zum Autofahren*, 2, verbesserte und erweiterte Auflage, Wien, Köln and Weimar.

Knoflacher, H. (1997), *Landschaft ohne Autobahnen: für eine zukunftsorientierte Verkehrsplanung*, Böhlau Verlag, Wien, Köln and Weimar.

Knoflacher, H. et al. (1995), 'Sustainable Development – Öko-City. Projektgruppe 1: Mobilität in der Stadt', Durchgeführt im Auftrag der Wiener Internationalen Zukunftskonferenz (WIZK).

Mumford, L. (1984), 'Die Stadt', DTV Wissenschaft No. 4326, München.

Newman, P.W.G. and Kenworthy, J.R. (1989), *Cities and Automobile Dependence: A Sourcebook*, Gower, Essex.

Schafer, A. (1998), 'The Global Demand for Motorized Mobility', *Transportation Research Part A*, 32 (6), pp. 455–77.

Wortmann, W. (1985), 'Wandel und Kontinuität der Leitvorstellungen in der Stadt- und Regionalplanung', *Berichte zur Raumforschung und Raumplanung*, 29 (3–4), pp. 20–25.

Chapter 18

Sketch Planning Transport Models: A Tool to Forecast Transport Demand in Eastern Europe

Paul Pfaffenbichler and Günter Emberger

Introduction

Several Central and Eastern European countries (CEEC) will join the European Union in the near future. The accession process challenges politicians and planners in Eastern and Western Europe. The region along the Austrian border, with its northeastern neighbours Czech Republic, Slovakia and Hungary (Figure 18.1), will be affected by fundamental changes. The transformation of former planning market economies into free market economies intensifies the exchange of goods and services. Omitting customs formalities and border waiting times will affect freight and passenger traffic. Residents of the border region expect negative consequences to result from the increased level of passenger and freight traffic volumes. Likewise, they will also expect politicians to respond promptly to these threats. Appropriate measures have to be put into place beforehand. A prerequisite to be efficient is the knowledge of the system reactions. Predictions of effects have to be made. Transport models can be used as a support tool for this purpose.

Long term predictions of economic and socio-demographic changes are always uncertain. Fundamental changes like the accession process increase the level of uncertainty furthermore. The simulation of a great number of potential scenarios is a way to cope with such a situation. But this approach requires short model run times and availability of appropriate data to set up the model. Today's standard four-stage[1] transport models are not fully suitable to meet this requirement. Reasons for this are:

- to model a detailed road and rail network with all attributes needed in the assignment stage is costly in terms of time and money;
- although computational power increases steadily, the run time for the

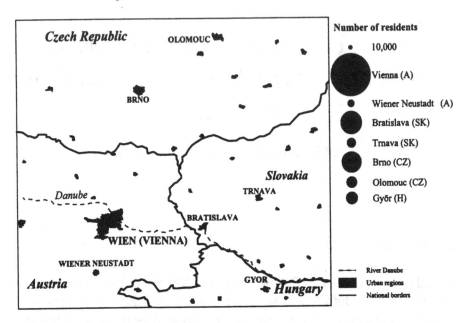

Figure 18.1 **Definition of the study area: border region Austria, Czech Republic, Slovakia, Hungary**

assignment of a detailed network is still too long, e.g., a run time of one hour to test 100 alternatives requires a computational time of more than four days;

- a calculation of intra-zonal traffic is not possible. This requires a fine-meshed zoning and increases the effort to collect data and to set up the model as well as the run time of the model.

Therefore the Institute for Transport Planning and Traffic Engineering, University of Technology Vienna (TUW-IVV[2]) develops and applies so called 'sketch planning models'. These models are based on a three-stage approach.[3] The assignment stage is deliberately left out. This allows very short run times. Sketch planning transport models have been used in several completed European research projects[4] and there is still further development in ongoing European research projects.[5]

The aim of this chapter is to demonstrate that sketch planning models are suitable tools to support planning processes in the context of the enlargement of the European Union.

The Modular TUW-IVV Decision Support Tool

Sketch planning models and decision support tools developed by TUW-IVV have been presented at several international conferences.[6] This chapter gives a rough summary and refers to the literature. Figure 18.2 shows an overview of decision-making in reality and how the TUW-IVV tool simulates these processes. In reality, stakeholders define targets. The decision-makers give weight to the targets and compare the weighted result with the reality. Based on this comparison decisions are made. The TUW-IVV tool simulates this reality in a modular way. Module 1, the transport model itself, simulates the behaviour of the transport system. Module 2 describes which policy instruments can be applied and how they affect the transport system. Policy measures could be, for example, infrastructure investments, changes in frequency and fare changes for railways and road capacity, road pricing and parking charges for car. Module 3, a set of objective functions, transforms real life targets into mathematical functions based on transport model output. Module 4, the optimisation method, tries to find the policy measure combination, which gives the maximum value for a certain objective function.

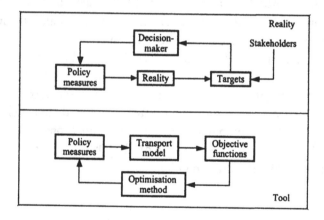

Figure 18.2   Decision-making process in reality and its simulation with a tool (Emberger, 1998)

Transport Model

The underlying principle for all strategic transport models created at TUW-IVV is the analogy to the law of gravity or to Kirchhoff's law[7]. The model uses simultaneous trip distribution and mode choice algorithms (Equation 1).

$$T_{ijm} = \sum_p P_{ip} * \frac{A_{jp} * f(t_{ijmp})}{\sum_{kl} A_{kp} * f(t_{iklp})}$$ **Equation 1**

where:

T_{ijm} number of trips from i to j by mode m;
P_{ip} production in zone i for purpose p;
A_{jp} attraction in zone j for purpose p;
t_{ijmp} impedance from i to j by mode m for purpose p; and
$f(t_{ijmp})$ friction factor from i to j by mode m for purpose p.

Computerised models of the Austrian cities Eisenstadt and Vienna have been created in the project OPTIMA. In the subsequent project, FATIMA, these models were developed further. The changes affected transport modelling itself and the introduction of new objectives into the cost benefit analysis. In OPTIMA and FATIMA only passenger transport and the purposes 'commuting' and 'non working' were considered. The modes pedestrian, bicycle, private car and public transport were taken into account.

A sketch planning model of Europe was developed in the project SAMI. It is based on the same principles as the previous urban transport models. Freight transport and the modes air, inland waterways and short sea shipping were added. The intra-zonal traffic is simulated in five distance classes. The model subdivides the geographical area of Europe into 9 zones and is given the acronym EURO9.[8] A detailed description of EURO9 is given in Pfaffenbichler and Emberger, 2000 and Pfaffenbichler, 2000.

Policy Instruments

On the demand side the implementation of policy instruments affects either travel time or travel costs. On the supply side the implementation of policy instruments causes investment and operating costs. Policy instruments can be discrete or continuous. They can be spatial homogenous or unhomogenous distributed. They can be applied for only a certain periods of time or for the whole day. Table 18.1 shows the policy instruments modelled in OPTIMA and FATIMA. In SAMI different levels of the policy measures could be applied in the categories cost and capacity in each zone, for each mode and purpose. This results in a maximum number of 113 independent measures.[9]

Table 18.1   Policy instruments modelled in the projects OPTIMA and FATIMA

Public transport	Infrastructure investment (discrete: none, low, high)
	Frequencies (continuous)
	Fares (continuous)
Private car	Road capacity[*] (continuous)
	Road pricing (continuous, only in certain areas)
	Parking charges (continuous, distinction between long and short term)

*   Soft measures like optimising traffic signals, telematics, etc.

Objective Functions

Objective functions are mathematical formulations of targets. The core of the objective functions used in OPTIMA, FATIMA and SAMI is a cost benefit analysis. A weighted average of today's and future costs and benefits is calculated to reflect the target of sustainability. Additional targets like reduction in emissions, accidents, etc. are considered by applying penalties and constraints.[10]

Optimisation Method

Different mathematical methods can be used to find an optimum. In OPTIMA and FATIMA a statistics-based method was used, while in SAMI an automated operations research routine was used. But the analysis of a number of different scenarios could be seen as an optimisation method as well. In the context of the accession of Eastern European countries to the European Union many decision-makers with different targets are involved. In SAMI a method named 'decentralised optimisation' was used to cope with competing objectives. Different objective functions reflect different targets, e.g., the focus can be more on economic or more on sustainability aspects. The optimisation method tries to find an optimal solution with regard to the different targets.[11]

Development of a New Sketch Planning Model

Defining the Task

A strategic transport should be used to give a first quantitative assessment of changes in transport caused by the accession to the European Union. The spatial dimension of the models used in the projects OPTIMA and FATIMA was the city. In the project SAMI the whole territory of Europe was modelled. The task addressed here is situated between these two applications (Figure 18.1). The limited availability of resources justifies the exclusion of freight traffic. Only passenger traffic is considered in the model. In the model the accession to the European Union is in a first approximation considered as omitting the border waiting time in road traffic. Rail, water and air traffic do not change. Other effects like the free movement of labour and housing or changes in land use could not be considered at the current stage.

Is it Possible to Use the Previously Developed Model EURO9?

Concerning the accession of the Czech Republic, Slovakia and Hungary to the European Union the EURO9 zones Central (C) and Eastern (E) Europe have to be considered. Zone C includes Austria, Belgium, Germany, Liechtenstein, Luxembourg, the Netherlands and Switzerland. Zone E includes Croatia, the Czech Republic, Hungary, Poland, Slovakia and Slovenia. Source and destination of the modelled inter-zonal traffic are Cologne and Budapest.

Due to the decrease in customs inspections the road mileage per year increased by about 150 million person kilometres (pkm). Both in rail and air transport the mileage decreased by about 30 million pkm. The total passenger mileage increased by 90 million pkm. The number of trips in road passenger transport increased by 121,000 per year. Both in rail and air traffic the number of trips decreased by about 40,000. The total number of trips between zone C and E increased by 41,000. As the trip distance is constant the percentage changes are the same for pkm and number of trips. Road traffic increases by +1.3 per cent, rail and air traffic decreases by –0.5 per cent and total inter-zonal traffic increases by +0.3 per cent.

The small number of trips is caused by the size of the zones and therefore the big distance between their centres of gravity. Effects which happen mainly in the border region of the zones, like the case here, can not be reflected by the model. EURO9 is therefore not applicable to the task addressed in this chapter.

Design of a Tailor-made Sketch Planning Model

As EURO9 turned out to be unsuitable, a new sketch planning model was designed. This model is tailor-made to predict the effects of the European Union accession for the region of Eastern Austria. It incorporates the cities Vienna, Wiener Neustadt, Györ, Bratislava, Trnova, Brno and Olomouc (Figure 18.1). The gravity model uses friction factors as defined in Walther et al. (1997). The intra-urban traffic is modelled in five distance classes. This approach is similar to the modelling of the intra-zonal traffic in EURO9 (Pfaffenbichler and Emberger, 2000). The modes pedestrian, bike, public transport and private car are modelled. The intra-urban traffic forms a reservoir for destination changes. Travel times of inter-urban public transport come from the timetable of the Austrian federal railway lines (ÖBB, 2001). Travel times of inter-urban transport by private car were calculated based on distances measured from a car map.[12] Modal split percentages from an Austrian study and average daily traffic volumes of border crossing points were used for provisional calibration (Snizek and Rosinak, 1998; BMwA, 1998; ÖSTAT, 1999).

The fall of customs barriers could be seen either as a policy instrument or as an external factor affecting reality/transport model. This decision is not in the hand of a single decision-maker. Therefore it is more appropriate to define it as an external factor. The modelled policy instruments are: mileage dependent road pricing, speeding-up the railway lines and public transport fare reduction.

As an example of a political objective, zero growth in road traffic volumes due to the European Union accession is chosen.

Results of the Sketch Planning Model Calculations

In the model prediction for the current situation about 35,000 trips per day cross the Austrian border with the Czech Republic, Slovakia and Hungary. This represents about 0.4 per cent of the total number of trips in the study area. The share of rail for trips crossing the border is about 10 per cent. An Austrian study gives for the rail traffic between Vienna and Bratislava a comparable share of about 8 per cent (Snizek and Rosinak, 1998).

If only the fall of custom formalities is considered in the model, the total number of trips in the relevant corridors will increase to about 42,000 per day of which about 38,500 are road trips (Figure 18.3, scenario 1). The number of trips by rail will stay nearly constant. Most increases will take place on the road. About 34 per cent of the additional road trips come from destination changes in car traffic. About 21 per cent will result from a shift away from public transport. About 45 per cent of the additional road trips are substituting non-motorised intra-urban trips. The road traffic in the whole study area increases by about 86 million pkm. Rail traffic decreases by about one million pkm. Considering a car occupancy rate of 1.4 persons per car this is an increase of about 61 million vehicle kilometres.

In a second scenario the European Union principle of free movement of labour is considered. An Austrian study (Huber, 2001) calculates the number of daily commuters from the Czech Republic, Slovakia and Hungary into the Vienna area with about 30,000. This number includes an interim period with partial restrictions on the movement of labour. In this case the total number of border crossing trips will increase to about 60,000 per day of which about 54,000 are road trips (Figure 18.3, scenario 2). These are about 0.6 per cent of all trips made in the region. The number of rail trips will increase by about 2,000. The number of car trips will increase by about 19,000. About 43 per cent of the additional road trips come from destination changes in car traffic. About 8 per cent will result from a shift away from public transport. About 49 per cent of the additional road trips are substituting non-motorised intra-urban trips. The road traffic will increase by about 420 million pkm. Rail traffic increase by about 50 million pkm. Considering a car occupancy rate of 1.4 persons per car this is an increase of about 300 million vehicle kilometres.

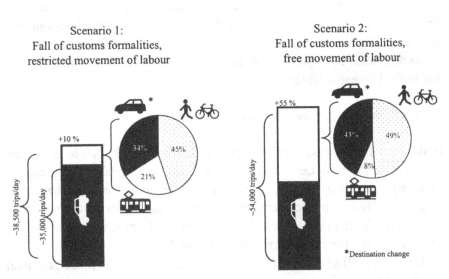

Figure 18.3   Calculated effects of accession to the European Union on the border crossing road corridors

Which measures must be implemented to achieve the given goal? For zero growth in road traffic volumes a road pricing on the border crossing corridors of about €0.1 per vehicle kilometre would be necessary in scenario 1. In scenario 2 road pricing of about €0.4 per vehicle kilometre would be necessary. This is about equal to doubling the cost per kilometre. Changes in the public transport system only have minor effects on road traffic volumes. For example, in scenario 1 a fare reduction of 50 per cent only reduces the increase in road traffic by about 1 per cent. However, the number of public transport trips in the corridor increases (+4 per cent). Measures favouring public transport result more in a shift in public transport destination (23 per cent) and from non-motorised modes (57 per cent) than in changes from car (20 per cent).

Conclusions

If overall effects are the subject, the 'sketch planning model' concept is suitable to forecast transport demand in the context of Eastern European countries accession to the European Union. Reasons are:

- the requirements of quality and level of spatial disaggregation for the underlying data are fulfilled by routinely gathered statistical data. Therefore data are available in the accession countries as well as in the concerned countries of the European Union. The costs to get access to these data can be neglected;
- the expenditure to apply a sketch planning model is lower as for common four-stage models;
- the 'sketch planning model' concept is able to cope with high level of uncertainties (equal or even better than a more detailed model[13]);
- the very short run time of 'sketch planning models' allows the simulation of high numbers of different potential economic, demographic, etc. scenarios;
- the 'sketch planning model' is able to deal with competing objectives in a multiple player environment.[14]

A prototype of a tailor-made sketch planning model was created. Interviews with transport experts at the University of Technology Vienna verified that the model behaves plausibly. The model described here was already used to calculate approximations for the changes in CO_2 emissions due to different scenarios for the European Union enlargement.[15] Nevertheless it has to be kept in mind that the 'sketch planning model' concept is not suitable for predicting local effects like noise beside a certain road, accident focus points, etc. Data required to calibrate the model, unlike basic data to set up the model, are not easily available. Existing data about medium and long distance travel do not entirely fulfil calibration requirements. Therefore some gaps remain in the model calibration. Expertise and resources from ongoing and future research projects will be used to improve the database for calibration. For example, TUW-IVV will be responsible for inter-urban transport in the future project SPECTRUM.[16] The sketch planning model presented here will form the basis of this work. Extending the model by freight transport will be an important issue. Extensions in the underlying theory are also proposed. In the ongoing project PROSPECTS a land-use submodel is added to an urban sketch planning transport model. These results should also be incorporated into a future regional model.

Notes

1   The four stages are: demand, distribution, mode choice and assignment. See, for example, Knoflacher et. al., 2000, p. 565.
2   Technische Universität Wien, Institut für Verkehrsplanung und Verkehrstechnik.
3   In 1980 the three-stage transport model ORIENT was presented by the University of Karlsruhe (Sparmann, 1980). But the proposed purpose for this model was to calculate input data for an assignment model.
4   4th FTE Framework: OPTIMA: Optimisation of Policies for Transport Integration in Metropolitan Areas; FATIMA: Financial Assistance for Transport Integration in Metropolitan Areas; SAMI: Strategic Assessment Methodology for the Interaction of Common Transport Policy Instruments.
5   5th FTE Framework: PROSPECTS: Procedures for Recommending Optimal Sustainable Planning of European City Transport Systems.
6   Knoflacher et al., 2000; Pfaffenbichler, 2000; Pfaffenbichler and Emberger, 2000 and Pfaffenbichler and Emberger, 2001.
7   Knoflacher et al., 2000; Pfaffenbichler, 2000; Pfaffenbichler and Emberger, 2000 and Pfaffenbichler and Emberger, 2001.
8   The model EURO9 was assessed against a multi-modal, network-based transport model of the European Union which was developed in the project STREAMS (Strategic Transport Research for European Member States). Despite fundamental differences in the spatial coverage, the zoning system and the representation of modes the principal model behaviour was the same.
9   Knoflacher et. al., 2000; May, 2000 and Pfaffenbichler and Emberger, 2001.
10 May et al, 2000; Knoflacher et al., 2000 and May et al., 2001.
11 Emberger, 1998 and Knoflacher et al., 2000.

12 Freytag and Berndt, *Autokarte Mitteleuropa*, 1:2 000 000, ISBN 3-85084-216-9.
13 See Knoflacher et al., 2000, p. 566
14 SAMI Final Report, 2000..
15 Rauh et al., 2001.
16 The project SPECTRUM (Study of Policies regarding Economic instruments Complementing Transport Regulation and the Undertaking of physical Measures) startes in January 2002.

References

BMwA (1998), *Automatische Straßenverkehrszählung 1998*, ed. Bundesministerium für wirtschaftliche Angelegenheiten, Band 1, Wien.

Emberger, G. (1998), 'Vorstellung einer Methode zum Lösen komplexer Optimierungs-probleme', in Manfred Schrenk (ed.), *Proceedings CORP 1998: Computergestützte Raumplanung*, Institut für EDV-gestützte Methoden in Architektur und Raumplanung, Wien, pp. 305–13.

Huber, P. (2001), 'Migration und Pendeln infolge der European Union-Erweiterung (Teilprojekt 10)', PREPARITY – Strukturpolitik und Raumplanung in den Regionen an der mitteleuropäischen European Union-Außengrenze zur Vorbereitung auf die European Union-Osterweiterung, Studie des Österreichischen Instituts für Wirtschaftsforschung im Rahmen der Gemeinschaftsinitiative INTERREG IIC, Wien.

Knoflacher, H., Pfaffenbichler, P. and Emberger, G. (2000), 'A Strategic Transport Model-based Tool to Support Urban Decision-making Processes', in J.-C. Mangin and M. Miramond (eds), *Proceedings 2nd International Conference on Decision-making in Urban and Civil Engineering*, 1, INSA Lyon (Fr), ESIGC Chambery (Fr), ENTPE Vaulx-en-Velin (Fr), ETS Montreal (Ca), Lyon, pp. 563–74.

May, A.D. (2000), 'From Problems to Solutions', *Integrating Transport in the City – Reconciling the Economic, Social and Environmental Dimensions*, OECD Proceedings, Paris, pp. 19–28.

May, A.D., Shepherd, S.P., Minken, H., Markussen, T., Emberger, G. and Pfaffenbichler, P. (2001), 'The Use of Response Surfaces in Specifying Transport Strategies', *Transport Policy*, 8, pp. 267–78.

May, A.D., Shepherd, S.P. and Timms, P.M. (2000), 'Optimal Transport Strategies for European Cities', *Transportation Research A*, pp. 285–315.

ÖBB (2001), 'Web page Österreichische Bundesbahnen', Timetable, <http://www.oebb.at>, accessed 22 May.

ÖSTAT (1999), 'Strassenverkehrszählung 1995 – Bundesstrassen im gesamten Bundesgebiet der Republik Österreich', ed. Österreichisches Statistisches Zentralamt, *Beiträge zur Österreichischen Statistik*, 1.319. Heft, Wien.

Pfaffenbichler, P. (2000), 'Ein strategisches Verkehrsmodell von Europa (EURO9)', ed. H. Knoflacher, Institut für Verkehrsplanung und Verkehrstechnik, Symposium 'Donauraum – European Union-Osterweiterung', Wien.

Pfaffenbichler, P. and Emberger, G. (2000), 'Ein strategisches Modell von Europa (EURO9)', *Proceedings CORP 2000: Computergestützte Raumplanung*, Ed. Manfred Schrenk; Institut für EDV gestützte Methoden in Architektur und Raumplanung; Wien, pp. 273–9.

Pfaffenbichler, P. and Emberger, G. (2001), 'Ein strategisches Flächennutzungs-/Verkehrsmodell als Werkzeug raumrelevanter Planungen', in Manfred Schrenk (ed.), *Proceedings CORP 2001: Computergestützte Raumplanung*, Institut für EDV-gestützte Methoden in Architektur und Raumplanung, TU Wien, Wien, pp. 195–200.

Rauh, W., Stögner, R., Kromp-Kolb, H. and Pfaffenbichler, P. (2001), 'Klimafaktor Vekehr – Wege zur klimaverträglichen Mobilität', Wissenschaft and Verkehr, vol. 4., ed. VCÖ, Wien.

SAMI (2000), 'Final Report – Guide for Strategic Assessment on CTP-Issues', Strategic Assessment Methodology for the Interaction of CTP-Instruments, project funded by the European Commission under the Transport RTD Programme of the 4th Framework Programme.

Snizek, S. and Rosinak, W. (1998), 'Verkehrskonzept Nordostraum Wien', Erstellt im Auftrag der Länder Burgenland, Niederösterreich, Wien im Rahmen der Planungsgemeinschaft Ost, Regionalconsulting Ziviltechniker GesmbH, Wien.

Sparmann, U. (1980), 'ORIENT – Ein verhaltensorientiertes Simulationsmodell zur Verkehrsprognose', in W. Leutzbach (ed.), *Schriftenreihe des Instituts für Verkehrswesen der Universität Karlsruhe*, Heft 20, Karlsruhe.

Walther, K., Oetting, A. and Vallée, D. (1997), 'Simultane Modellstruktur für die Personenverkehrsplanung auf der Basis eines neuen Verkehrswiderstands', in W. Schwanhäuser and P. Wolf (eds), *Veröffentlichungen des Verkehrswissenschaftlichen Instituts der Rheinisch-Westfälischen Technischen Hochschule Aachen*, Heft 52, Aachen.

Chapter 19

Key Problems with Dry Cargo Handling in the Principal Bulgarian Ports of Varna and Bourgas

Boyan Kavalov

Introduction

Over the past few years several key transformations have taken place in the world bulk carriers' shipping segment. These changes have been caused mainly by the qualitative development and significant improvement of port infrastructure all over the world. It became possible for larger ships to call for full and complete loading and discharging at a significantly increased number of ports. Both loading and discharging operations have been significantly accelerated. As a result of all these factors, the average size of operated fleet has risen noticeably, as shown in Table 19.1.

The exploitation of these larger ships, together with their improved handling in the ports, has led to a significant decrease in transport costs, due to the increased economy of scale. Operating Handy Size ships has become a less attractive alternative, in comparison with Handymax vessels. In reality, the Handymax ships have now started to replace the Handy Size fleet. Additionally, there was another substantial qualitative increase of the average size of vessels, because of the intensive rise in the Panamax sector. At present, the economy of scale ensuing from exploitation of bigger bulk carriers is expected to reflect in a noticeable decrease in the corresponding world freight rates.

For these reasons it is of great importance for each country to identify how its existing national seaport infrastructure meets these newly-emerged trends. It is also highly recommended that the main existing or anticipated obstacles at the seaport network are identified, which might otherwise prevent the complete and profitable utilisation of benefits from these trends in global sea transportation.

Table 19.1 Development of world bulk carriers' fleet by class in selected months within the period 1997–2001 – number of vessels and total deadweight (dwt)

Vessel's class	Handy size (10000–30000 dwt)		Handymax (30000–50000 dwt)		Panamax (50000–80000 dwt)	
Month/year	Number	Million dwt	Number	Million dwt	Number	Million dwt
March 1997	2241	49.2	1725	67.6	933	61.2
March 1998	2269	49.6	1774	70.1	990	65.5
March 1999	2206	48.3	1773	70.6	1007	67.1
March 2000	2150	47.1	1789	71.4	1039	69.9
January 2001*	2111	46.1	1801	72.0	1070	72.3
March 2001	2101	45.9	1903	77.4	991	68.5

* As from February 2001 vessels up to 55000 dwt are included in the Handymax section. Vessels between 55000 dwt and 80000 dwt are included in the Panamax section.

Source: Adapted from 'The Drewry Monthly' of Drewry Shipping Consultants Ltd.

Current Status of Port Infrastructure for Dry Cargo Handling in Bulgaria

Bulgaria possesses two main Black Sea port complexes, situated around the towns of Varna and Bourgas. These two ports handle not only the great majority of the country's annual waterborne dry cargo turnover, but also the main share of its overall seaway import and export freight transport flows, as shown in Table 19.2.

Table 19.2 Bulgarian seaport freight transport flows within the period 1995–99 (thousand tonnes)

Bulgarian seaport cargo flows	1995	1996	1997	1998	1999
Total cargo passed	14416	21536	20668	17846	15848
Import (discharged)	5697	13344	11695	10237	9154
Export (loaded)	7958	7908	8241	6936	6602
Transit	761	284	759	669	78
Coastal	–	–	–	4	14

Source: National Statistical Institute of the Republic of Bulgaria.

The Port Complex of Varna

The port complex of Varna comprises four harbours handling dry cargo: three public (Varna-East, Varna-West and the port of Balchik) and one private (the coal harbour of the thermal power station 'Devnia', operated by the National Electric Company). The key technical parameters of these four harbours are given in Table 19.3.

Varna-East handles general cargo, various chemicals, metals, grain, fodder, bulk sugar, kaolin, etc. There are electric cranes up to 32 tonnes and one floating crane up to 100 tonnes.

The coal harbour of the thermal power station 'Devnia' is linked to the open sea by a navigable channel (Channel No. 1). Besides the draft restriction in this channel, the access to the harbour aquatoria is also limited by air-draft due to the Asparouh's bridge over the channel. Cranes of up to 40 tonnes handle the imported coal.

Table 19.3   Key technical parameters of harbours in the port complex of Varna, handling dry cargo

Port of Varna	Varna-East	Coal Harbour	Varna-West	Balchik
Number of berths*	13	3	18	1
Total quay length	2415 m	511 m	3100 m	164 m
Draft in vessel's berths	23–36 ft	28–37 ft	28–33 ft	23 ft
Air-draft restriction	–	41.72 m	41.72 m	–
Draft in connecting channels	–	37 ft	30 ft	–
Handling vessels up to	50000 dwt	50000 dwt	30000 dwt	5000 dwt
Covered storage facilities	39142 m²	–	37000 m²	–
Open-air storage facilities	107370 m²	5000 m²	347000 m²	14000 m²

*   For handling dry cargo only.

Table 19.4   Key technical parameters of harbours in the port complex of Bourgas, handling dry cargo

Port of Bourgas	East Harbour	Bulk Harbour	West Harbour
Number of berths*	13	5	5
Total quay length	1965 m	750 m	890 m.
Water draft in the vessel's berths	24–33 ft	36 ft	36 ft
Handling vessels up to	25000 dwt	44000 dwt	40000 dwt
Covered storage facilities	44500 m²	5000 m²	11000 m²
Open-air storage facilities	50000 m²	49000 m²	191000 m²

*   For handling dry cargo only.

Varna-West, situated along the Beloslav Lake, is linked to the open sea by a second channel (Channel No. 2).[1] This harbour specialises in handling cement, soda, clinker, coal, coke, fertilisers, ores and ore concentrates, quartz sand, etc. The dry cargo are handled by cranes up to 35 tonnes.

The Port of Balchik, situated 45 km north of Varna handles grain and is equipped with two cranes of 10 tonnes each.

The Port Complex of Bourgas

The port complex of Bourgas includes three public harbours for handling dry cargo. Their key technical parameters are given in Table 19.4.

The East Harbour specialises in handling general cargo, non-ferrous metals, foodstuff, bagged salt, fertilisers, bentonite, etc. Berths are equipped with cranes of 16 tonnes. The Bulk Harbour handles raw materials, exclusively coal, ores, concentrates, clinker, etc. There is a handling facility specifically for coal, with 1,200 tonnes discharging capacity per hour. The West Harbour handles various kinds of metals and dry cargo. Berths are equipped with cranes up to 40 tonnes.

Key Problems with the Bulgarian Port Infrastructure for Handling Dry Cargo

Draft Restrictions at the Ports of Varna and Bourgas

Due to the existing draft limitations in both Varna and Bourgas ports, bulk carriers with deadweight of more than 50,000 tonnes cannot call at any Bulgarian port for full and complete loading or discharging. This rules out the most widely-used Handymax and Panamax classes. Moreover, the access to the main Varna harbour for dry cargo handling – Varna-West – is considerably impeded by the existing tight combination between air-draft restriction in Channel No. 1 and tighter draft limitation in Channel No. 2. In addition, the draft limitations in both channels are not constant, because of rapid silting. For instance, the draft restriction in Channel No. 2 varies between 29 and 31.6 ft, depending on the time lapse since the last dredging. As a result of this unfavourable series of navigation obstacles, even the passing of ships of low Handymax class via Channel No. 2 becomes complicated. These vessels are assessed individually on their ability to navigate the channel, based on their specific technical data and construction parameters.

From the above, it can be summarised that the extent of draft restrictions governing access to both Varna and Bourgas ports does not correspond to the prevailing status and expected trends in world-wide shipbuilding and operating. The inability to accept the most widespread bulk carriers of upper Handymax class, as well all Panamax vessels for full and complete loading and discharging leads to significant loss of economy of scale. The losses caused by the need to use Handy Size and Handymax vessels fall into two categories: firstly, there is lost economy of scale, due to the relative increase of the daily time-charter hire – per tonne dwt or per tonne cargo on board. As proof, some calculations are given in Tables 19.5 and 19.6. Secondly, there are losses in economy of scale in fuel costs, due to the relative increase of fuel consumption of intermediate fuel oil (IFO) and marine diesel oil (MDO) – per tonne dwt or per tonne cargo aboard.

Table 19.5   Daily time-charter rates for different classes of bulk carriers (built 1985–90) basis: 12-month time-charter contracts, prompt delivery, within the first quarter of 2001

Vessel's class	Handy Size	Handymax	Handymax	Panamax
Deadweight	26000–28000	35000–37000	40000–45000	64000–65000
January 2001	US$ 6750	US$ 7500	US$ 8000	US$ 10000
February 2001	US$ 6500	US$ 7250	US$ 7700	US$ 10000
March 2001	US$ 6750	US$ 7500	US$ 8250	US$ 9750

Source: Adapted from 'The Drewry Monthly' of Drewry Shipping Consultants Ltd.

Table 19.6   Time-charter hire per tonne dwt for average bulk carrier of different classes (built 1985–90) basis: 12-month time-charter contract, prompt delivery, within the first quarter of 2001

Average deadweight*	27000	36000	42500	64500
January 2001	US$ 0.25	US$ 0.21	US$ 0.19	US$ 0.16
February 2001	US$ 0.24	US$ 0.20	US$ 0.18	US$ 0.16
March 2001	US$ 0.25	US$ 0.21	US$ 0.19	US$ 0.15
Average difference with the previous class		–US$ 0.04	–US$ 0.02	–US$ 0.03

*   Average vessel's deadweight is calculated basis ship's dimension range by classes bulk carriers, taken from the second row of Table 19.5.

It is not feasible to calculate the actual extent of savings in the economy of scale from fuel consumption, due to the variety in the daily fuel consumption by types of ships' engines. The direct economy of scale from fuel costs noticeably rises in importance in periods of high world prices of fuels, for example during the period since the end of the first quarter of 2000, as shown in Table 19.7.

Due to the port restrictions in Varna and Bourgas, the larger vessels of upper Handymax and Panamax class can only enter or leave their aquatoria in partial loaded condition. However, this constraint considerably increases freight rates and final delivery cost. These big bulk carriers are forced to call at a second port for additional loading or preliminary discharging. In both cases payment of extra disbursement costs at the second loading or at the first discharging port results in the increase of sea transport prime cost, freight rates and final delivery value. Another opportunity – additional loading or preliminary discharging on roads – cannot currently be considered, as dry cargo handling in both Varna and Bourgas ports cannot be operated in this mode. There are various projects aiming to improve the Bulgarian port infrastructure to overcome the disadvantages from the restrictions at Varna and Bourgas. The most important of these are for the development of the Bourgas port complex.

The Master Plan for developing the port of Bourgas by 2015 envisages under-stage and parallel building of 4000 metres of quay walls divided into four specialised terminals, two of them for handling dry cargo.

First of all, a harbour (Terminal No. 1) is planned for handling general and liquid cargo. It will consist of four berths of 750 m total length. The annual handling capacity of general cargo in this new harbour is expected to reach 0.35 million tonnes, with 0.7 million tonnes of covered storage capacity available per annum.

The second new harbour (Terminal No. 2) will be destined for handling bulk dry cargo and metals exclusively. It will include six berths of 1,580 m total length, constructed in three stages. It will have a final handling capacity of 3.4 million tonnes coal, 1.5 million tonnes ores and 1 million tonnes general cargo. The technical parameters of the berths will allow handling of large bulk carriers with deadweight up to 100,000 tonnes, 260–270 m LOA and draft between 42 and 51 ft.

Low Loading and Discharging Rates in Varna and Bourgas for Some Dry Cargo

At Varna and Bourgas ports, the loading and discharging rates for some dry cargo are lower than the rates in other Black Sea ports, especially those of the

Table 19.7 Average March bunker prices of IFO and MDO in selected major bunkering ports within the period 1997–2001 (US$ per tonne)

Area Port Year	Arabian Gulf Mina Al Ahmadi		Mediterranean Genoa		NW Europe Rotterdam		US Gulf Houston		Far East Singapore	
	IFO	MDO	IFO	MDO	IFO	MDO	IFO	MDO	IFO	MDO
1997	102	217	105	185	94	152	96	182	98	189
1998	74	170	81	145	72	123	64	152	71	120
1999	66	125	77	132	64	101	67	110	66	98
2000	171	263	171	233	154	222	145	238	174	230
2001	120	260	142	245	129	200	132	240	136	195

Source: Adapted from 'The Drewry Monthly' of Drewry Shipping Consultants Ltd.

Community of Independent States (CIS). As proof, Table 19.8 presents the usual loading rate for a typical dry cargo, shipped out ex-Black Sea – urea fertiliser. Furthermore, the resulting adverse impacts on the time-charter equivalent[2] and seaway prime cost are shown as well. The upward trend in the time-charter hire when loading urea in Varna-West does not take into consideration that the loading in the Bulgarian port is performed on the basis of SHEX UU[3] terms, but not SHINC[4] terms, as it is in the Yuzhny[5] fertiliser terminal.

Poor Quality of Loading Operations in Bulgarian Ports

Loading of packed dry cargo: bags, big-bags, sling-bags, etc. Principally, the rate of the ship's bale capacity is calculated on the basis of conscientious and careful placement of the full cargo into all vessels' holds. When packed goods are loaded in Bulgarian ports, gangs are sometimes careless about the quality of stowing operations. Goods are directly dumped into ships' holds, without enough or no further arrangement or stowing. This incurs damage of goods, injuries to vessels' holds and increases the corresponding stowage factor (SF) above its usual rate. This increase reflects in a decrease of the total cargo taken aboard. Referring to the time-charter calculation approach, these negative factors lead to a noticeable increase of freight rates.

Loading of light [(SF) above 40 cubic feet] dry bulk cargo When light goods are intensively and quickly loaded or when they are loaded in relatively small shiplots in Bulgarian ports, usually only natural trimming takes place, with no additional mechanical trimming. In this case, light goods do not have enough time for satisfactory natural trimming and consolidation. As a result, vessels load less cargo than expected. This means inefficient utilisation of ships' cargo grain capacity and increased sea transport prime cost, i.e., freight rates.

Lack of Adequate Terrestrial Network, Connecting Port of Balchik

The main reason for the poor turnover in the port of Balchik is the lack of an appropriate and well-developed connecting overland infrastructure.

The current status and condition of the road infrastructure are definitely unsatisfactory. Moreover, problems with good and suitable road connections with the Varna port have been significantly complicated over the past few years by the heavy landslide activity north of Varna.

There is no railway link to the Balchik port. The absence of rail transport connections with the local cargo suppliers and receivers significantly

Table 19.8 Comparative increase in the freight rate for shipment of bulk urea in 20,000 tonnes shiplot ex-Varna-West/Bulgaria, in comparison with 20,000, 30,000, 40,000 and 60,000 tonnes shiplots ex-port of Yuzhny/Ukraine, basis actual loading rates, due to the lower loading rate in Varna-West and the respective increase in the time-charter (t/c) hire period, calculated at March 2001

Parameters/Ports	Varna-West	Yuzhny			
1 Shiplot's quantity (tonnes)	20000	20000	30000	40000	60000
2 Loading rate (tonnes per day)	5000	20000	20000	20000	20000
3 Days in loading [(1)÷(2)]	4	1	1.5	2	3
4 Time-charter rate (US$ per day)*	6750	6750	7500	8250	9750
5 Total T/C hire (US$), [(3)×(4)]	27000	6750	11250	16500	29250
6 T/C hire (US$ per tonne), [(5)÷(1)]	1.35	0.34	0.38	0.41	0.49
7 Difference Varna-West/Yuzhny (US$ per tonne)	1.01	0.97	0.94	0.86	

* Time-charter rates for the corresponding to the shiplots bulk carriers are taken from the last row of Table 19.5.

discourages freight transport flows through the port. As a result, the available port infrastructure is not utilised, which means no recovery of investments and fixed costs. Furthermore, the under-utilisation of Balchik is a significant loss of real benefits, as its average draft limitation enables use by bulk carriers for short sea navigation.

Conclusions

The inability of both the main Bulgarian ports of Varna and Bourgas to fully and completely handle larger bulk carriers of upper Handymax and Panamax classes results in a significant increase in the sea transport prime costs, freight rates, total transport costs and final import and export delivery prices. The approach for evaluating this lost economy of scale, calculated by goods and destinations,[6] is given in Table 19.9.

Table 19.9   Approach for evaluating the lost economy of scale on FOB and CFR parities and in freight rates, due to the utilisation of Handy Size (HS) bulk carriers, in comparison with Panamax (P) ships

Parameters/vessel's class	Handy Size	Panamax
1 FOB price for the goods assessed	US$/tonne	US$/tonne
2 Voyage-charter freight rate	US$/tonne$_{hs}$	US$/tonne$_p$
3 Final delivery CFR price $[(1)+(2)]$	US$/tonne$_{hs}$	US$/tonne$_p$
4 Absolute surplus in the freight rate $[(2)_{hs}-(2)_p]$	US$/ton	–
5 Relative surplus in the freight $[(4)\times100\div(2)_p]$	%	–
6 Relative surplus in the FOB price $[(4)\times100\div(1)]$	%	–
7 Relative surplus in the CFR price $[(4)\times100\div(3)_p]$	%	–

The actual extent of this loss of economy of scale for the Bulgarian economy cannot be evaluated with a satisfactory accuracy, because it varies considerably with time, depending on the following key factors:

- ongoing structural transformations in the Bulgarian economy and especially in its industrial subsector. This directly reflects into both aggregate demand and aggregate supply, and respectively into the outgoing and incoming seaway cargo flows;

- dynamic and variable foreign trade policy. Duty rates represent one of the most important factors, affect the export and import cargo flows, usually vary within the year and cannot be foreseen;
- the clearly stated 'open' nature of the Bulgarian economy: with insufficient energy resources of its own, the state of the Bulgarian economy is inversely related to the worldwide market and prices of the main raw materials – oil, natural gas, etc.

From the above, the losses of scale in shipping for the Bulgarian economy depend on the following parameters:

- nominal and real production capacity of plants by sectors of industry;
- actual productivity of plants by sectors of industry;
- degree of export/import orientation of plants by sectors of industry;
- nature of goods shipped in/shipped out by destinations;
- typical ships' rotation(s), by goods or by types of goods assessed – loading and discharging area(s), port(s), etc.;
- freight rates by products, destinations and other specific factors.

However, it may be concluded that this loss of economy of scale determines the competitiveness when transport of mass cargo is included, i.e., when goods have relatively low production value and freight rates for their transportation are a considerable share of the final delivery cost.

The problems to the Bulgarian economy ensuing from the national port infrastructure reflect a decrease of both regional – within the Black Sea area – and worldwide export competitiveness of the country. There are also additional internal costs due to the increased import price level. These disadvantages increase in impacts within periods of downward trends in the global economy, distinguished generally by depressed aggregate demand and low world prices of commodities.

Notes

1 Vessels with length over all (LOA) exceeding 200 m, beam over 26 m and gross register tonnage (GRT) over 20,000, are required to use Channels No. 1 and No. 2 during daylight hours only.
2 Usual approach for calculating sea transport prime cost, based on the rate of the daily time-charter hire.
3 Saturdays, holidays excluded unless used.

4   Saturdays, holidays included.
5   Main CIS export fertiliser terminal, being part of the Odessa port complex.
6   Price parities as per INCOTERMS.

References

Branch, A.E. (1995), *Economics of Shipping Practice and Management*, Chapman and Hall, London.

Chernomorie Daily (1995), *Bulgarian Marine Industry 95*, Popular Deed Press, Varna.

Drewry Shipping Consultants Ltd. (1995), *Panamax Bulk Carriers – Market Prospects and Profitability, 1995–2000*, Drewry Shipping Consultants Ltd, London.

Drewry Shipping Consultants Ltd. (1997), *The Drewry Monthly – April 1997*, Drewry Shipping Consultants Ltd, London.

Drewry Shipping Consultants Ltd. (1999), *The Drewry Monthly – April 1999*, Drewry Shipping Consultants Ltd, London.

Drewry Shipping Consultants Ltd. (2000), *The Drewry Monthly – January 2000*, Drewry Shipping Consultants Ltd, London.

Drewry Shipping Consultants Ltd. (2001), *The Drewry Monthly – April 2001*, Drewry Shipping Consultants Ltd, London.

National Statistical Institute (2001), *Statistical Yearbook 2000*, Statistical Press and Print, Sofia.

Index

Printed in the United States
by Baker & Taylor Publisher Services